CONCORD
PUBLICATIONS COMPANY

by CONCORD PUBLICATIONS CO.
603-609 Castle Peak Road
Kong Nam Industrial Building
10/F, B1, Tsuen Wan
New Territories, Hong Kong
www.concord-publications.com

We welcome authors who can help expand our range of books. If you would like to submit material, please feel free to contact us.

We are always on the look-out for new, unpublished photos for this series. If you have photos or slides or information you feel may be useful to future volumes, please send them to us for possible future publication. Full photo credits will be given upon publication.

ISBN 962-361-122-6
printed in Hong Kong

FENNEK

Desert Foxes in Germany

The new reconnaissance vehicle Fennek

Ralph Zwilling

The 177kW Deutz 6-cylinder BF6M 2013C diesel engine - with its 5.7-liter displacement, exhaust turbocharger, intercooler and electronic fuel injection - provides a maximum speed of 115km/h. However, the maximum speed is limited to 90km/h by order of the German Department of Transportation. Note the antenna mount and the Eiserne Kreuz (Iron Cross) marking on the side of the hull.

Development and Evaluation

German Bundeswehr reconnaissance troops have never been blessed with the latest equipment and the Spähpanzer 8x8 Luchs, introduced in 1975, became obsolete even after some major improvements like the addition of a thermal sight and a new navigation system. It was therefore high time that a completely new armored, wheeled reconnaissance vehicle be developed and procured. Handover of the first recon vehicle (Spähwagen) Fennek to the Bundeswehr at the German Armor School on 10 December 2003 can therefore be viewed as a milestone in procurement of equipment for the Aufklärungstruppe. Initially the Spähpanzer Luchs 2A2 will be augmented, before being completely replaced, by the Spähwagen Fennek. Like its namesake, an agile desert fox with large ears, the Spähwagen Fennek conducts surveillance and reconnaissance missions.

As the German Panzeraufklärungstruppe was never blessed with the latest equipment, the handover of the first Fennek reconnaissance vehicle to the Bundeswehr at the German Armor School in Munster on 10 December 2003 can be viewed as a milestone. Like its namesake, an agile desert fox with large ears, the Fennek is extremely capable at conducting surveillance and reconnaissance missions.

The impressive engine power output of 29.5kW per cylinder is accomplished by a four valves per cylinder layout, combined with a magnetic valve system developed by Deutz, and a high-pressure fuel injection system. The Deutz diesel engine, located in the middle at the rear of the vehicle, meets the emission levels specified by Euronorm III. Note the short ground sensor equipment antenna on the rear beside the smoke dischargers.

The Fennek has add-on all-round armor protection against 7.62mm Armor-Piercing (AP) rounds (Dragunov Standard), a level accomplished by the installation of thicker armored glass in the windows and composite armor plates on the outside of the vehicle. Depending on mission requirements, the vehicle can easily be equipped with different add-on armor protection made from rolled homogeneous steel or different composite materials.

In contrast to the old reconnaissance vehicle Spähpanzer Luchs 2A2, the Fennek cannot swim, but is able to cross water obstacles up to 100cm deep without the need for any preparation. The Spähfahrzeug can also negotiate 60% forward and 35% side slopes. Note the electrically controlled and heated rear-view mirrors which can be easily removed if necessary.

The drive train features a fully automatic transmission connected to a transfer case providing selectable four-wheel drive to a limited-slip differential. The H-layout of the drive train permits a lower silhouette, with the power being transmitted from the transfer case to the wheels via angular wheel gears. Note the driver's hatch in the front while the commander's and system operator's hatches are parallel to each other in the rear.

Originally the project started through the initiative of the General of the German Armor Troops, Major General Grillmeier, on 1 August 1988. By 10 October 1988, the tactical requirements were forwarded to the appropriate agencies in the German Department of Defence and evaluation trials were conducted between the Panhard VLB and the Zobel, produced by the Gesellschaft für Systemtechnik. The search for a future wheeled recon vehicle started in 1992, with a presentation of different concepts by the military industry. The Netherlands joined the project in 1993 and since the Koninklijke Landmacht planned to procure more vehicles than the German Heer, it played the leading role from the beginning of the project. The German and Dutch militaries have a long tradition of cooperation; working together to develop the Armored Recovery Vehicle 3 Büffel based on the Leopard 2, and the Air Defence System Gepard based on the Leopard 1. From the beginning, the Fennek and its three-man crew were developed to conduct reconnaissance missions as a two-vehicle section behind enemy lines for a period of at least five days. The vehicle required sufficient storage capacity to enable the crew to transport the necessary equipment, supplies, water and food to accomplish assigned missions. Requirements also specified an airlift capability aboard both the C-160 Transall and C-130 Hercules cargo airplanes. Furthermore, the vehicle dimensions had to permit transport by rail, trucks and ships according to NATO standards.

In the end, the German/Dutch concept (a joint venture between DAF Special Products Aerospace & Vehicle Systems (SP) and Wegmann) won the bilateral evaluation. SP was responsible for the chassis while Wegmann

A central tire inflation system enables the driver to adjust tire pressure on the move from his control panel, according to the prevailing terrain conditions. The Michelin XML 12.00 R20 tires, with run-flat inserts, can run for 100km even after being damaged by bullets or shrapnel. The cross-country performance of the Fennek is outstanding.

Since the driver's vision is limited by a blind spot behind the vehicle when reversing, there is a camera attached to the rear of the upper hull with a black and white monitor located to the right of the steering wheel for the driver.

While the African predator has to rely on its eyes and ears, the German Fennek has the latest technical equipment. Key items include an all-weather observation and reconnaissance unit located behind the commander's station, state-of-the-art command and control equipment, and a central system control panel. The LITEL LLN-GY hybrid navigation system, composed of inertial and global positioning system (GPS) units, permits accurate determination of the vehicle's position and direction with respect to grid north.

On 21 December 2001, a contract was signed for the combined procurement of 612 Fennek vehicles. The German Heer ordered 178 reconnaissance and 24 combat engineer versions while the Koninklijke Landmacht will procure 202 reconnaissance, 130 MRAT (Medium Range Anti-Tank) and 78 general purpose versions. Delivery of all 612 Fennek will be accomplished by 2007. Note the armored window on the side of the hull providing the commander with an outstanding view to the side.

assumed responsibility for the remaining components. In December 1994 the Dutch Directie Materieel and the Joint Venture SP/Wegmann signed a contract for production and delivery of two Dutch and two German prototypes with an option for future series production. The first Dutch prototype, named TVM 1 (Truppenversuchmuster 1 / Troop Trail Model 1), was handed over to the Koninklijke Landmacht on 18 December 1996. From 1996 to 1997, all four prototypes (TVM 1-4) were delivered and tested during technical trials conducted by the German Administration for Defence Technology and Procurement (BWB - Bundesamt für Wehrtechnik und Beschaffung) and its counterpart in the Dutch

From the beginning, the Fennek with its three-man crew was developed to conduct reconnaissance missions as a two-vehicle section behind enemy lines for at least five days. The front of the Fennek is also used to store four snow chains, seven 5-liter jerry cans, four tow pintles, hand tools, and the camouflage panels to cover the windows and the front lights.

The observation and reconnaissance unit named Beobachtungs – und Aufklärungsausstattung (BAA) comprises an OPHELIOS thermal imager, a CCD day vision camera and a MOLEM laser rangefinder. Note the Rotzler Treibmatic TR 30/5 cable winch mounted on the front of the Fennek which can be used for self-recovery. The length of the 13mm thick winch cable is 46 meters.

Depending on mission requirements, the Fennek can also be equipped with a 7.62mm x 51 MG3 machine gun instead of the 40mm x 53 Grenade Machine Gun (GMG). Access to the crew compartment is provided by two large doors on the sides of the chassis and three hatches in the upper hull.

With torsion bar suspension, shock absorbers and ebonite bump stops, the 10.65-ton Fennek has outstanding cross-country mobility even in difficult terrain. Its strengths lie in its high on- and off-road mobility. In such terrain the High Mobility Multipurpose Wheeled Vehicle (HMMWV) would probably get stuck immediately. Seven windshield wipers are used for cleaning the three large armored windows at the front of the vehicle.

For observation from concealed positions, the compact sensor head can be tripod-mounted at an exposed location for remote operation up to a distance of 37m from the vehicle. Note the wiper used to clean the BAA lenses, and the Fennek in the background. It only takes five to ten minutes to install the remote controlled observation and reconnaissance unit on a tripod since the entire sensor can be removed from the telescoping mast mount on top of the Fennek.

By using the telescoping mast-mounted sensor head, the Fennek can stay concealed behind hedges or small hills. Both the observation and reconnaissance unit, and the weapon station, are remote controlled under armor. With the help of the thermal sight, targets can be detected from a distance of up to 7,000 meters, even at night or during times of limited visibility. Definite target identification can be achieved from a distance closer than 2,500 meters.

Department of Defence. Three tactical trials were conducted between September 1997 and August 1998 at several army bases in Germany and the Netherlands, these being led by both the German Armor School in Munster and the Dutch OTCMAN in Amersfoort. During these tactical trials the prototypes and crews had to participate in numerous five-day reconnaissance missions and complete 35 day and night recon courses travelling as far as 120km. Obviously the cross-country performance and mobility of the new vehicle was tested under realistic conditions and the results were positively convincing.

While testing the four TVM prototypes, both Armies recognized that some of the original requirements had to be modified or were no longer necessary. SP/Wegmann was asked to edit the concept and present its results no later than June 1997. Changes to the concept specification resulted in an additional prototype based on the TVM 2, and was therefore named TVM 2 mod. This TVM 2 mod., delivered on 2 July 1999, included all developmental changes made on the other four prototypes during

The fuel tank, situated in the rear of the vehicle close to the engine compartment, has a capacity of 236 liters, which provides a cruising range of up to 1,000km on roads and 500km cross-country. Note the white Tarnkreuz found on many Bundeswehr vehicles and that is used for convoy operations at night or in low visibility conditions. The recon companies of the Bundeswehr intervention units (Eingreifkräfte – EK) will have three platoons with two recon sections, each equipped with two Spähwagen.

This photo shows the upper hull with the driver's hatch in the front. The commander's and system operator's stations are parallel to each other in the rear of the crew compartment. Note the anti-slip coating and the KMW Type 1530 weapon station equipped with the new 40mm Heckler & Koch Grenade Machine Gun.

technical and tactical trials. From 2 July 1999, all tests were conducted using the TVM 2 mod. During this period, SP/Wegmann built another reference system (RS) at the firm's own expense for further testing and optimization. Due to the merging of various German defence companies in recent years, the Fennek is now built by Krauss-Maffei Wegmann GmbH & Co. KG (KMW), together with Dutch Defence Vehicle Systems (DDVS). DDVS is a 100% subsidiary of KMW. On 21 December 2001, a contract was signed for the combined procurement of 612 Fennek vehicles: 202 reconnaissance, 130 MRAT (Medium Range Anti-Tank) and 78 general purpose versions for the Royal Netherlands Army; 178

To reduce the Fennek's infrared signature, the engine compartment is equipped with an advanced cooling and ducting system. The air intake in the front, covered with wire mesh, is used to cool the Renk 606 automatic transmission. The round fan in the middle of the upper hull draws hot air out of the engine compartment. The hatch on the left provides access to the NBC overpressure system, and also includes the fresh air intake for cooling the engine, the engine sockets and the starter. The opening on the left rear is the air-intake for the seven radiators of the Fennek. The hatch on the right rear provides access to the stowage box holding the BAA tripod, three ground sensor units and batteries, three sleeping bags, 7.62mm and 40mm ammunition boxes, a spare barrel and spare breech for the MG3 machine gun.

The heart of the Fennek is the proven and technically advanced observation and reconnaissance unit named Beobachtungs – und Aufklärungsausstattung (BAA). The BAA is comprised of an OPHELIOS thermal imager, a CCD day vision camera and a MOLEM laser rangefinder. If necessary, the entire unit can be folded and stored under armor in the upper hull. Note the round fuel filler cap left of the BAA.

All observation equipment is packaged in a sensor head mounted on a telescoping mast that is situated behind the commander's station. Using a small joystick on the BAA control unit, the sensor head can be controlled +/- 220° in azimuth and +/- 30° in elevation. It can also be raised to a height of 3.29m above the ground (i.e. 1.5m above the vehicle's roof). The commander's hatch is located in front of the telescoping mast.

Only one Fennek in each Recon section (Spähtrupp) is equipped with a cable drum. This drum is located in a storage box in the rear, with the cable being used for tripod-mounted remote controlled operations up to 37 meters from the vehicle. Note the smoke grenade launcher system installed on the rear, rear-view camera in the middle, and the armor plates attached to the chassis.

For self-defence all German Fenneks are primarily armed with the new 40mm grenade machine gun (GMG) produced by Heckler and Koch. The weapon, which can be operated under armor and NBC protection, has an effective range of up to 600 meters. The sight system consists of a PERI Z17 B with an uncooled thermal sight. Additionally, the system operator has two normal periscopes through which to view the battlefield. There are 32 ready-to-fire 40mm x 53 grenades available for the GMG and another 32 stored in a stowage box in the rear of the upper hull.

To clean the lenses of the observation and reconnaissance unit, there is a water-jet system incorporated into the hatch behind the mast mount.

Depending on the mission, the crew can easily replace the GMG with the reliable 7.62mm x 51 MG3 machine gun. With the MG3, 120 or 250 ready-to-fire 7.62mm rounds are available while another seven ammo boxes, each with 120 rounds (or four boxes each with 250 rounds), are stored in the rear of the upper hull.

reconnaissance and 24 combat engineer versions for the German Army. The delivery of all 612 Fenneks will be accomplished by 2007. As a rapid fielding initiative, the German Army ordered and has already fielded four Fenneks as artillery forward observer versions serving with the German ISAF contingent in Afghanistan.

A Tour through the Fennek

The technical performance of the new Fennek is amazing. Its strengths lie in its high on- and off-road mobility, which is achieved by using state-of-the-art engine technology. The 177kW Deutz 6-cylinder diesel engine BF6M 2013C, with its 5.7-liter displacement, exhaust turbocharger, intercooler and electronic fuel injection, provides a maximum speed of 115km/h. However, the maximum speed is limited to 90km/h by order of the German Department of Transportation. The impressive power output of 29.5kW per cylinder is accomplished by a four valves per cylinder layout combined with a magnetic valve system developed by Deutz, and a high-pressure fuel injection system. The BF6M 2013C engine, located in the middle of the rear of the vehicle, meets the emission levels specified by Euronorm III.

The drive train features a fully automatic Renk 606 transmission with six forward and one reverse gear connected to a transfer case providing selectable four-wheel drive to a limited slip differential. The H-layout of the drive train permits a lower silhouette, with the power being transmitted from the transfer case to the wheels via angular wheel gears. With torsion bar suspension, shock absorbers and ebonite bump stops, the 10.65-ton Fennek has outstanding cross-country mobility even in difficult terrain. The fuel tank, situated in the rear of the vehicle close to the engine compartment, has a capacity of 236 liters, which provides a cruising range of up to 1,000km on roads and 500km when traveling cross-country. In contrast to the old reconnaissance vehicle Luchs 2A2, the Fennek cannot swim, but is able to cross water obstacles up to 100cm deep without preparation. Furthermore, the vehicle can negotiate 60% forward and 35% side slopes. A central tire inflation system enables the driver to adjust tire pressure while on the move according to the terrain conditions. The Michelin XML 12.00 R 20 tires with run-flat inserts can run for a distance of up to 100km even after receiving bullet or shrapnel damage. Since the driver cannot see in the blind spot behind the vehicle while reversing, there is a camera attached to the rear of the upper hull, with a black and white monitor located to the right of the driver's steering wheel.

The Fennek has add-on all-round protection against 7.62mm Armor-Piercing (AP) rounds (Dragunov Standard). This is accomplished by

Access to the 177kW Deutz 6-cylinder diesel engine is provided by a large hatch in the middle of the rear. Left of this hatch is the storage box for the BAA cable drum. Before the fumes leave the exhaust system located on the right side of the vehicle's rear, they are cooled by the exhaust air from the engine radiators. Besides two radiators for the engine, the Fennek has a radiator for transmission oil, hydraulic fluid, fuel, turbocharger air, and the recuperator of the air-conditioning system. Note the rear-view camera and the six smoke grenade launchers.

The commander's station is located on the left side behind the driver. All crew member stations are fitted with well-designed seats equipped with seatbelts. They were optimized to meet ergonomic requirements, and are designed in accordance with German road licensing regulations. Note the bracket system in front of the seat and the smoke grenade launcher system device attached to the roof interior close to the hatch.

The inside of the crew compartment is covered with a special foam material designed to prevent the formation of water condensation. The bags attached to the door can hold hand grenades, maps and other personal equipment of the crew. Note the two door handles, hinges and grounding strap.

installing thicker armored glass windows and composite armor plates on the outside of the vehicle. Depending on mission requirements, the vehicle can easily be equipped with different add-on armor protection made from rolled homogeneous steel or different composite materials. The crew compartment is protected against anti-personnel mines containing charges of up to 5kg of TNT. To offer the crew maximum survivability, all equipment such as radios, control panels and monitors are mounted on shock absorbing mounts. The infrared signature is minimised through special exhaust ducting. Before the fumes leave the exhaust system, they are cooled by exhaust air from the engine radiators. Besides two radiators for the engine, the Fennek has a radiator for the transmission oil, hydraulic fluid, fuel, turbocharger air, and the recuperator of the air-conditioning system. Equipped with an advanced nuclear, biological and chemical (NBC) warfare protection system that is integrated into the air-conditioning system of the crew compartment, the Fennek can also operate in contaminated environments.

The Fennek's driver sits in the front of the crew compartment with an excellent 180° view through the armored glass windows. Both the commander and the system operator sit parallel to each other in the rear behind the driver. The driver's controls and displays were developed in accordance with ergonomic requirements, thus making it easy for the driver to handle the large number of vehicle functions. Left of the steering wheel are the controls for the Rotzler Treibmatic TR 30/5 cable winch mounted in the front of the Fennek, and the Sygeon central tire inflation system. Located on the right side of the instrument panel are the controls for the rear-view camera monitor, four-wheel drive, differentials and the gear lever. The Fennek is not equipped with a fire extinguisher system, but rather with only a fire

The height adjustment for the commander's and the system operator's ISRI seats are operated by an electric motor which permits both under-armor and out-of-the-hatch observation. In the event of a sudden emergency, the observer can engage a quick lowering mechanism and withdraw into the crew compartment within seconds. During rest phases, it is possible to relax under the raised seats, a position providing some comfort during five-day recon missions.

warning system. For personal protection, the driver is armed with a 9mm x 19 pistol P8 produced by Heckler & Koch in Germany. The use of LUCIE image intensifier goggles makes it easier for the driver to safely operate the vehicle at night or during periods of low visibility. Produced by Thales, the LUCIE goggles can be used for 60 to 80 hours depending on the batteries being used. For a 5-day reconnaissance mission, each Fennek crew has fifteen meals, ready to eat (MREs), and approximately 60 liters of water.

All crewmember stations are fitted with comfortable seats equipped with seatbelts. They were optimized to meet ergonomic requirements and are designed in accordance with German road licensing regulations. The height adjustment for the commander's and the system operator's ISRI seats is operated by an electric motor which permits both under-armor and out-the-hatch observation. Additionally, both seats are equipped with 4-point suspender seatbelts and can be swivelled 360°. In the event of a sudden emergency, the observer can engage a quick lowering mechanism and withdraw into the crew compartment within seconds. During rest phases, it is possible to relax under the raised seats to provide some comfort during 5-day recon missions. Normally the commander of the Fennek monitors the battlefield from the open hatch. The diameter of his hatch is 8cm larger than the hatch on other Bundeswehr combat vehicles, allowing him to comfortably observe while wearing the new body armor of the "Infanterist der Zukunft" (Infantry Soldier of the Future).

The bracket system includes the panel for the observation and reconnaissance unit, the command and control equipment and the central system control panel. Depending on the mission, this bracket can be easily moved with a hand-wheel located between the commander's and the system operator's stations.

The position of the system operator is parallel to the commander in the rear of the crew compartment. During missions he normally operates the observation and reconnaissance unit, command and control equipment and central control panel. He is also responsible for the ground sensor equipment when it is in operation.

The system operator is also in charge of the KMW Type 1530 weapon station including a gun mount designed to carry the Grenade Machine Gun or the 7.62mm MG3 Machine Gun. Note the sight system consisting of a PERI Z17 B with 4x magnification, an uncooled thermal sight and two periscopes.

While the African predator relies on its eyes and ears, the Spähwagen has the latest technical equipment. Key items include an all-weather observation and reconnaissance unit, command and control equipment and a central system control panel, all of which are mounted in a multipurpose bracket system. Depending on the mission, this bracket can be easily moved with a hand-wheel between the commander's and the system operator's stations.

The heart of the Fennek is the technically advanced and proven observation and reconnaissance unit named Beobachtungs – und Aufklärungsausstattung (BAA) - which comprises an OPHELIOS thermal imager, a CCD day vision camera and a MOLEM laser rangefinder, all built by the German company Carl Zeiss Optronic GmbH. All observation equipment is packaged in a sensor head mounted on a telescoping mast situated behind the commander's station. Using a small joystick on the BAA control unit, the sensor head can be controlled +/- 220° in azimuth and +/- 30° in elevation. It can also be raised to a height of 3.29m above ground (i.e. 1.5m above the vehicle roof). For observation from a concealed position, the compact sensor head can be tripod-mounted at an exposed location for remote operation up to 37m from the vehicle. Only one Fennek in each Recon section (Spähtrupp) is equipped with a cable drum located in a storage box in the rear. By using the black and white monitor of the BAA control unit, located on the right side of the bracket system, the operator can switch between the thermal sight and CCD camera views. With the help of the OPHELIOS thermal sight, targets can be detected up to a distance of 7,000 meters, even at night or during limited visibility conditions. Definite target identification can be achieved from a distance closer than 2,500 meters. The BAA control unit also includes the controls for brightness, contrast, focus, washing system for the sensor head lenses, and several other menu functions.

The radio equipment consists of a SEM 80 (VHF), a SEM 90 (VHF), a HRM 7000 (HF) and a handheld SEM 52SL (VHF). These units are stored in the rear of the crew compartment between the commander's and the system operator's seats.

This view shows the bracket system to advantage. The command and control equipment (Führungskomponente - FüKom), located on the top left of the bracket system, integrates the reconnaissance vehicle into the C^3 network. By using the black and white monitor of the BAA control unit, located on the right side of the bracket system, the operator can switch between the thermal sight and CCD camera views. Among other things, the central system control panel located below the FüKom monitor is used to control the telescoping mast, hybrid navigation system, ground sensor unit, and the built-in test system.

The command and control equipment (Führungskomponente), which is located on the top left of the bracket system, integrates the reconnaissance vehicle into the C^3 network. The FüKom consists of a computer, 15" touch screen and keyboard. For orientation in cross-country terrain, digitized maps are used showing the vehicle's position, target position, and current

The Fennek driver sits in the front of the crew compartment with an excellent 180° view through the armoured glass windows. His controls and displays were developed in accordance with ergonomic requirements making it easy for him to handle the large number of vehicle functions. Left of the steering wheel are to be found the controls for the front winch and the Sygeon central tire inflation system. Located on the right side of the instrument panel are the controls for the rear-view camera monitor, four-wheel drive, differentials and the gear lever.

The smoke screen of the anti-infrared smoke grenades fired by the Fennek prevents observation by enemy day/night vision devices and greatly interrupts acquisition by targeting devices and infrared or heat-seeking missiles. (PzTrpS Grp WE Dez 4)

battlefield environment. This information is continuously updated between vehicle and command post via data radio link with HF and VHF radio. One SEM 80 (VHF), one SEM 90 (VHF), one HRM 7000 (HF) and a handheld SEM 52SL (VHF) radio are stored in the rear of the crew compartment. Initially FüKom will be equipped with the "FüInfoSysHeer" software which is currently used by other Army troops (Arrmour, Infantry, Engineers etc). Projected software is the reconnaissance data, information and control network software (DIFA), which will be 80% FüInfoSysHeer and 20% troop specific adaptations. This will also include interfaces used to link the software with unmanned aerial vehicles (UAVs) and the new ground sensor units.

Among other things, the central system control panel located below the FüKom monitor is used to control the telescoping mast, hybrid navigation system, ground sensor unit, and the built-in test system. It can also be used to start the engine or to control the capacity of the batteries. The LITEL

Recently three Fenneks were tested in Saragossa, Spain with different camouflage systems. One of them was the Barracuda Mobile Camouflage System (MCS) produced by Saab from Sweden. MCS is designed to reduce heat penetration through the hull into vehicles that are operating in extremely hot climates like deserts and the tropics, in order to protect personnel and sensitive equipment. It utilizes different materials to maintain camouflage in the visual/near infrared, thermal, and radar wavebands. Further enhancements to the MCS include the application of heat transfer reduction materials to reduce the heating of the vehicle by solar radiation. (PzTrpS Grp WE Dez 4)

Requirements for the Fennek also specified an airlift capability aboard both the C-160 Transall and C-130 Hercules cargo airplanes. Detailed tests with prototypes were conducted at the Wunstorf Airfield near Hannover, Germany. (PzTrpS Grp WE Dez 4)

LLN-GY hybrid navigation system, composed of inertial and global positioning system (GPS) units, permits accurate determination of the vehicle's position and direction with respect to grid north. Determination of target coordinates is achieved using the system's laser rangefinder and azimuth and elevation measuring equipment, together with the navigation system. To locate a target, the operator has to push a button on the BAA control unit joystick. He can choose between the first and last laser echo. He then inputs target information into the command and control equipment. Information is easily displayed as an icon on the FüKom touch screen and transmitted with the data radios to other vehicles connected via FüInfoSysHeer or DIFA.

For self-defence, all German Fenneks are primarily armed with the new 40mm grenade machine gun (GMG) produced by Heckler and Koch. The weapon, which can be operated from under armor and NBC protection, has an effective range of up to 600 meters. The manually controlled but electrically fired KMW Type 1530 weapon station, located above the system operator's seat, includes a gun mount designed to carry the GMG or the 7.62mm MG3 machine gun. The sight system consists of a PERI Z17 B with 4x magnification and an uncooled thermal sight. Additionally,

Aladin is a mobile, compact, electrically-powered, miniature aerial vehicle (MAV). It consists of a backpack-mounted ground control station with data link, digital data recorder, extra lithium polymer battery and two 3kg airplanes, each of which has a 146cm wingspan. The MAV is launched either by hand or with a ground catapult powered by a bungee cord. In the air, Aladin operates autonomously on a pre-programmed flight route controlled by a digital autopilot coupled with a built-in GPS receiver. The route is planned with a touch-sensitive tablet computer combining both flight monitor and controls. (PzTrpS Grp WE Dez 4)

As a rapid fielding initiative, the German Army ordered and has already fielded four Fenneks as artillery forward observer versions serving with German ISAF troops in Afghanistan. This Fennek is armed with the 7.62mm MG3 machine gun. (Carl Schulze & Yves Debay)

the system operator has two normal periscopes through which to view the battlefield. There are 32 ready-to-fire 40mm x 53 grenades available for the GMG, and another 32 stored in a stowage box in the rear of the upper hull. These grenades include both HE and HEAT rounds in a ratio of 2:1. The GMG has a muzzle velocity of 248m/s and a rate of fire of up to 350 rounds per minute. The GMG is a very safe and versatile weapon that can fire both shrapnel and armor-piercing ammunition, thus making it the perfect main armament for the new Fennek. Depending on the mission, the crew can easily replace the GMG with the reliable 7.62mm x 51 MG3 machine gun. With the MG3, 120 or 250 ready-to-fire 7.62mm rounds are available, while another seven ammo boxes, each with 120 rounds (or four boxes each with 250 rounds), are stored in the rear of the upper hull. Both the commander and system operator are equipped with a 5.56mm x 45 G36A1 rifle produced by Heckler and Koch, and LUCIE night vision goggles. The smoke discharger system, mounted on the rear of the Fennek, creates a smoke screen approximately 50 meters in front of the vehicle within seconds.

Mobile and Stationary Reconnaissance Equipment

In addition to the observation and reconnaissance unit (Beobachtungs- und Aufklärungsausstattung - BAA), the Fennek recon sections will also be equipped with small unmanned aerial vehicles and ground sensor equipment.

Miniature Aerial Vehicle Aladin

Aladin is a mobile, compact, electrically-powered, miniature aerial vehicle (MAV). It consists of a backpack-mounted ground control station with data link, digital data recorder, extra lithium polymer battery, and two 3kg airplanes with a 146cm wingspan. The system can be transported by two soldiers and is ready for hand-launching in just five minutes. Aladin will be used for reconnaissance, observation, identification, locating and targeting. If necessary, the MAV can be equipped with a thermal sight. The MAV is launched either by hand or with a ground catapult powered by a bungee cord. In the air, Aladin operates autonomously on a pre-programmed flight route controlled by a digital autopilot coupled with a built-in GPS receiver. The route is planned with a touch-sensitive tablet computer combining both flight monitor and controls. If necessary, the operator can easily direct the flight using the "over-flight" or "circulate" functions to observe special terrain or targets. The video data is transmitted to the ground control station in real-time or recorded on miniature DV tapes. The 17kg ground control station comprises a 2D or 3D map for navigation and mission planning, this also possessing an obstacle avoidance function based on a 3D terrain model. The ground control station also includes a monitoring display, video display and digital recording module. Depending on the mission, the flight altitude ranges from 30 to 150 meters above ground level. The mission radius with Video/Telemetric Radio Link in real-time is up to 5km. With the MAV's maximum speed of 45 to 90km/h, it can operate for 30 to 45 minutes depending on weather conditions. To land, the aircraft cuts its engine several meters above the ground, pulls up into stall configuration and drops. After landing, the MAV is easily prepared for another mission. Preparation includes changing the batteries, two Styrofoam shock absorbers, and plastic tubes connecting the wings with the aircraft body. Currently the German Bundeswehr plans to purchase 115 Aladin MAV systems for its reconnaissance troops.

Ground Sensor Equipment

Stationary sensors like the ground sensor equipment (Bodensensorausstattung – BSA) produced by EXENSOR from Sweden, consist of six individual ground sensor units and enable roads and other focal points to be monitored. These sensors permit accurate detection, classification and type identification of combat and combat support vehicles, including their number, speed and direction of travel. Each recon section has one BSA with components carried on both vehicles. Each of the 12kg ground sensor units (Bodensenoreinheit - BSE) comprises a field computer connected via a cable with two multi-sensors. These sensors have an acoustic, seismic and 3-axial magnetic sensor. BSA allows reconnaissance troops to detect, classify and identify vehicles travelling on roads beyond visual range, as well as determining their direction and speed of travel. In cases where the BSA cannot positively type identify the target, it can, at a minimum, inform the Fennek operator of the vehicle class (light wheeled, heavy wheeled or tracked vehicle). Personnel and animals can also be detected when the operator selects a special local reconnaissance mode. In this mode the BSA can also detect targets with significant

Note the desert camouflage applied to this Spähwagen Fennek belonging to German ISAF troops seen near Kabul, Afghanistan. The crew also added a chevron made of white tape on top of the commander's hatch for the purpose of aerial identification. (Carl Schulze)

This Fennek took part in a search and destroy operation conducted in cooperation with the French Foreign Legion in the Shomali plain located 50km north of Kabul. (Yves Debay)

quantities of metal. Sensor data is collected, analysed and compared to an existing signature database in the field computer. Once target analysis is complete, the target is assigned a time stamp and unique identity and then transmitted to the Fennek that may be located up to 10km away.

The two camouflaged multi-sensors are deployed 10 meters away from a road or trail with a separation of 10 to 20 meters between sensors, with the field computer positioned in the middle. The optional motion sensor (PIR Sensor) has a range of up to 30 meters. If someone tries to remove the BSE or damage the cables, it sends a warning message to the Fennek central system control panel. The BSE can operate for about 14 days in a tactical environment and is virtually impervious to the weather. Two soldiers can deploy the sensors in less than fifteen minutes. The ground sensor equipment greatly enhances the efficiency of reconnaissance troops as all information detected is recorded in the field computer. In cases where the data-link between the Fennek and the BSE is interrupted, the operator can individually query each of the ground sensors and download data collected since communications were lost. Based on current information, the German Bundeswehr plans to purchase 95 Bodensensorausstattungen for its reconnaissance troops.

Future Organization of Bundeswehr Reconnaissance Battalions

As currently envisioned, the reconnaissance battalions of the German intervention units (Eingreifkräfte – EK) will consist of a headquarters and headquarters company, three reconnaissance companies, an operation and support company, a mixed UAV (Unmanned Aerial Vehicle) battery equipped with the system KZO and supported by a signal platoon. Each recon company will have three platoons with two recon sections each.

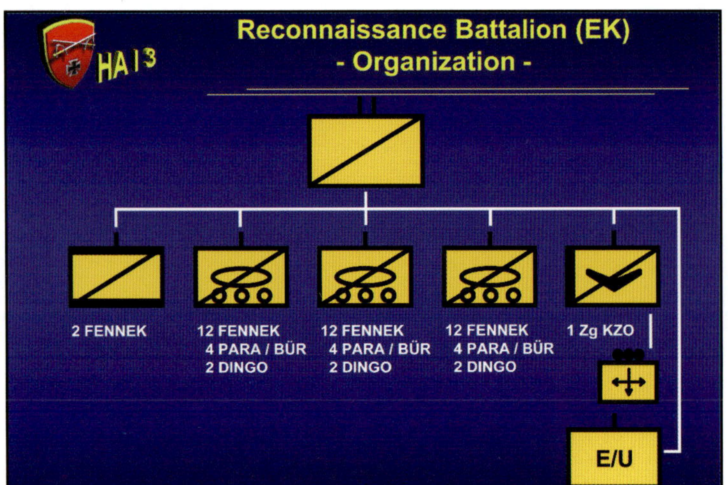

Especially for the ISAF mission, the German Bundeswehr Fennek artillery forward observer versions received an unique desert camouflage. The water soluble emulsion colors last approximately six months and can easily be removed with an emulsion and waterjet. Note the German flag attached to the antenna and the Eiserne Kreuz (Iron Cross) which is an identification marking for German Bundeswehr vehicles. (Yves Debay)

A Dutch Reconnaissance Company consists of three platoons each equipped with eight Fennek LVB (Licht Verkennings- en Bewakingsvoertuig), an anti-tank platoon with four Fennek MRAT (Medium Range Anti Tank) and a logistic platoon. The company headquarters will also be equipped with two Fennek AD (Algemene Diensttaken). (Yves Debay)

This Fennek LVB is equipped with louvers attached in front of the windows which should prevent the treasonable reflection of the sunlight during reconnaissance missions. (Yves Debay)

In contrast, the reconnaissance battalions of the stabilization units (Stabilisierungskräfte – SK) have a headquarters and headquarters company, two reconnaissance companies, an operation and support company, a mixed UAV battery equipped with the systems LUNA and KZO and a signal platoon. The SK battalions are expected to operate over a larger area of operations during stabilization missions and therefore, their recon companies have four platoons, each with two recon sections. One of the SK recon companies will be equipped with the well-known Spähpanzer Luchs 2A2 while the other company will have the Spähwagen Fennek.

The black mesh netting and the louvers both installed to prevent the treasonable reflection of the sunlight in detail. The black marking on the side of the vehicle is used as the base for the convoy identification number which is normally written with white chalk. (Yves Debay)

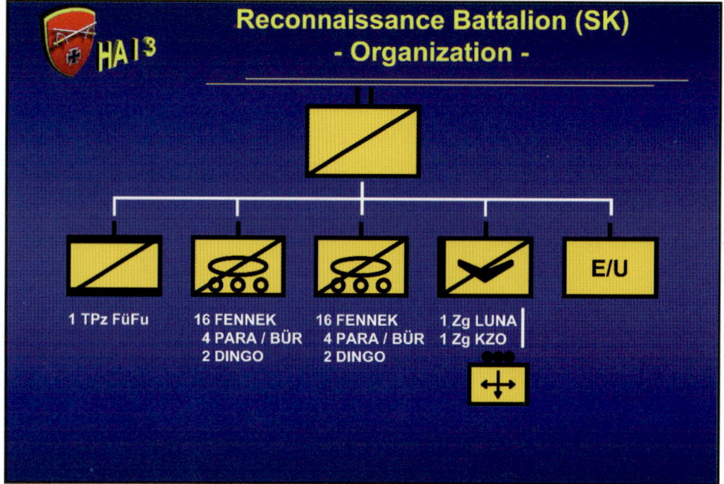

At the beginning the Royal Netherlands Army will get 202 reconnaissance, 130 MRAT (Medium Range Anti-Tank) and 78 general purpose versions of the Fennek. (Yves Debay)

A Fennek recon section in both the EK and SK companies consists of two vehicles. The commander of the 1st vehicle, normally a Master Sergeant (Hauptfeldwebel) is also the section leader. His system operator with the rank of a Sergeant First Class (Feldwebel), operates the weapons, reconnaissance and observation equipment, Aladin UAV and the ground sensor equipment. When the commander leaves the Fennek, the system operator is the executive section leader. The driver of the 1st vehicle, normally a Private First Class (Obergefreiter), is also a trained EOD specialist. While the 2nd vehicle is normally let by a Sergeant First Class, both its driver and system operator are enlisted soldiers.

Advanced observation and reconnaissance equipment in conjunction with new command, control and communication systems as well as a flexible weapon station with two different kinds of armaments enables the Fennek crew to conduct recon missions behind enemy lines and send information within seconds to command posts or other vehicles. With these capabilities, the small but agile Fennek outperforms other recon vehicles like the Spähpanzer Luchs 2A2 or even the M3A3 Bradley.

Acknowledgements
The author would like to thank Major Brune and Hauptfeldwebel Meyer-Urban from the German Panzertruppenschule in Munster, and Mr. Sharman of Krauss-Maffei Wegmann GmbH & Co. KG in Kassel, for their outstanding support.

In contrast to the German Spähwagen Fennek, the Dutch Fennek LVB version is armed with a cal.50 M2 machine gun. Note the Dutch markings at the front of this vehicle and the black mesh netting attached to the side window. (Yves Debay)

BATTLE GRIFFIN

Battle Griffin 2005

Three Block War in Ice and Snow

Carl Schulze

The CV 9040 has a crew of three, consisting of the driver, gunner and vehicle commander. In the vehicle's rear compartment, up to eight fully equipped infantrymen can be carried. A CV 9040 of I5 Regiment "Fältjägarna" can be seen preparing to go out on patrol. Note the vehicle's nickname painted in yellow on the side of the gun mount.

Protection for the Non-combatant Evacuation Operation (NEO) site was provided by heavily armed members of the Royal Netherlands Marine Corps. This Marine is armed with a 5.56mm C7A1 Diemaco assault rifle.

The ice-covered street is blocked by old oil drums, burning tires and pieces of wood. Around the barricade, groups of heavily armed militiamen can be seen keeping themselves warm by moving up and down the snowy sides of the road. Other groups of militiamen walk down the convoy which has come to a halt in front of the roadblock. The militiamen ask Multinational Task Force (MNTF) troops sitting in their vehicles for food, money, cigarettes or spirits. Meanwhile, at the head of the convoy, the Norwegian convoy commander tries to negotiate a way through for his convoy stuck at the roadblock. But the situation looks hopeless, as the militia leader and his men have previously been promised food and security for the Ranian enclave by the Multinational Task Force, but neither has materialized yet. Just then, CV 9030N infantry fighting vehicles of a Norwegian mechanized infantry company, and Bv 206 all-terrain vehicles of an Italian Alpini mountain infantry company, appear on the scene. These combat troops also belong to the Multinational Task Force, and have arrived on the scene in order to ensure freedom of movement for the MNTF convoy. After an apparent escalation, in which the militia soldiers prepare to engage the Norwegian and Italian troops, the situation calms a little when the militia commander recognizes that his force is totally outgunned. In the ensuing negotiations between the MNTF combat troop commander and the militia commander, a solution is finally reached. The road is opened for the MNTF convoy and the Italian Alpini mountain troops remain in the area to provide protection for the population of the enclave. What appears like a

In the outer security perimeter around the NEO site, the Netherlands Marines deployed anti-tank teams equipped with the "Gill" portable anti-tank weapon system. Just recently introduced in the RNLMC, Gill has a combat range of up to 2,500m. A complete Gill system has a weight of 26kg, of which the rocket weighs 13kg. Gill consists of the rocket, a tripod and the command and launch unit. The fire-and-forget anti-tank weapon system has a top attack capability, and is fully capable of night fighting. Gill was developed and manufactured by Rafael Armament DA in Israel.

scene from a NATO operation in Bosnia or Kosovo was in reality an incident during "Exercise Battle Griffin 2005" in Norway. Held between 21 February and 11 March 2005, the exercise involved 14,000 troops who were training for Crisis Response and Peace Support Operations under extreme winter conditions.

Battle Griffin 2005 - An invitational exercise

Exercise Battle Griffin 2005 was not a NATO exercise, but a Norwegian invitational exercise. In addition to NATO nations, the Norwegian government also invited Partnership for Peace (PfP) member nations to take part in the exercise. Battle Griffin 2005 was conducted as a live joint combined exercise (LIVEX) within the framework of a Deployed Forces Concept, aiming at encompassing all aspects of out-of-area operations. The aims of the Norwegian government for conducting Exercise Battle Griffin 2005 were as follows:

- Train for the deployment and employment of reaction forces into a crisis area to perform the full spectrum of Crisis Response Operations (CRO) under winter conditions.
- Improve operational skills at the tactical level in a NATO-supported CRO scenario, led by a National Joint Headquarters.
- Enhance interoperability between multinational force structures that can be a part of real world NATO and/or UN operations in harsh and cold climates.

The Norwegian government's invitation was extended to fourteen nations, all of which deployed troops to the exercise. In total, 14,000 troops of naval, ground and air forces participated in the exercise. Troops supplied by the invited nations were as follows:

On the pictured Bv 206 vehicle, mounted anti-tank teams of the 1st Battalion of the Royal Netherlands Marine Corps (RNLMC), provide the outer security ring for the NEO site. This team was deployed, together with a second one, along one of the main approach roads to the evacuation site. Note the .50-cal M2HB heavy machine gun and the "Gill" portable anti-tank weapon system placed on the roof of the Bv 206.

- Belgium: 100 troops with one Alouette III helicopter, four F-16 fighter aircraft, and the mine countermeasure ship "BNS Narcis".
- Denmark: 100 troops with six F-16 fighter aircraft and four Fennec helicopters. Some troops belonged to an Air Controller Unit.
- Finland: 700 troops of the Finnish Rapid Deployment Force, as well as two Mi-8 transport helicopters, and four MD500 liaison helicopters.
- France: 50 troops of an infantry platoon.
- Germany 1,600 troops with two UH-1D transport helicopters, twenty-one MRCA Tornado combat aircraft in different variants, and the mine countermeasure ship "FGS Dillingen".
- Italy: 200 troops of the 9th Alpini Regiment from L'Aouila. The mountain infantry troops belonged mainly to the 93rd Company.
- The Netherlands: 1,000 troops belonging to the 1st Battalion, The Royal Netherlands Marine Corps, and support elements. The Marines were deployed on board the Landing Platform Dock "HNLMS Rotterdam". Also deployed were two KC-130 tanker aircraft of the Royal Netherlands Air Force. The Dutch Navy took part with the frigate "HNLMS Karel Doorman", the mine countermeasure ship "HNLMS Urk", the supply ship "HNLMS Amsterdam", and the submarine "HNLMS Dolfijn".
- Poland: 100 troops with two MiG-29 fighter aircraft and one C295 M CASA transport aircraft.
- Romania: 8 troops acting as liaison officers.
- Spain: 100 troops with five F-18 fighter aircraft.
- Sweden: 700 troops with two CH-46 transport helicopters, two BO105 liaison helicopters, and eight JAS Saab Gripen fighter aircraft. The Swedish ground element consisted of a mechanized infantry company of

A crewmember of a Dutch Bv 206 all-terrain vehicle mans the vehicle's roof-mounted 12.7mm M2HB machine gun.

Dutch Marines use the light machine gun version of the 5.56mm C7 Diemaco at the section level. The weapon features a heavy barrel, a hydraulic recoil buffer, an adjustable bipod and a select fire mechanism. The weapon has a weight of 5.8kg, and a cyclic rate of fire of 625 rounds per minute. Here a Marine of the 1st Battalion of the Royal Netherlands Marine Corps provides cover for his comrades who are fully engaged in riot control duties.

Like several other nations, the Netherlands uses the Bv 206 all-terrain vehicle for operations in arctic climates and regions with deep snow. Among other tasks, the Bv 206 is used by the Dutch Marines as a tracked all-terrain personnel carrier. The Bv 206 has a weight of 4.49 tons, and is powered by a Mercedes Benz 2.996-liter six-cylinder diesel engine. The vehicle has a turning radius of 8m, and can reach a top speed of 52km/h. In the front cab there is space for up to six soldiers, while in the rear up to eleven can be transported.

I5 Regiment "Fältjägarna", based at Östersund, and equipped with ten CV 9040s, one CV 90 recovery vehicle, one MT-LB and one MT-LBU.

- Switzerland: 8 troops acting as observers.
- United Kingdom: The supply ship "RFA Sir Galahad", manned by members of the Royal Fleet Auxiliary.
- USA: 1,300 troops of Marine Air Ground Task Force 25 (MAGTF-25), and two KC-135 tankers of the USAF.

Land operations involved 7,000 troops from Norway, Italy, the United States, Finland, Sweden, Italy and France. At sea, forty-three ships and submarines were involved in the exercise with some 3,500 crewmembers. In the air, 104 aircraft and helicopters, along with 3,500 personnel, were involved in Exercise Battle Griffin 2005.

NATO units involved

Even though the exercise was a Norwegian one to which other nations were invited, multinational NATO units also took part. This included Combined Air Operations Center 3 (CAOC3), Standing Naval Mine Countermeasure Group 1 (SNMCMG1), the Multinational Electronic Warfare Signal Group (MEWSG), and aircraft of the Airborne Warning and Control System (AWACS).

Norwegian forces

Over 50% of the participating troops in Battle Griffin 2005 came from the host nation of Norway. Some 8,000 Norwegian troops of the Norwegian Army, Navy, Air Force and Home Guard were involved in the exercise. The bulk of the Norwegian ground forces that took part in the exercise belonged to the 6th Norwegian Division, which also acted as headquarters of the land forces of the MNTF. Norwegian ground forces included an armored battalion, a mechanized infantry battalion, the ISTAR (Intelligence, Surveillance, Target Acquisition, Reconnaissance) Battalion, engineer assets, combat service and support assets, and several Norwegian Home Guard units. Heavy equipment of the ground forces included Leopard 1A5NO main battle tanks, CV 9030N infantry fighting vehicles, the Leopard 1AVLB "Leguan", M113 armored personnel carrier variants (ambulance, engineer section vehicle, command post), Bv 206 all-terrain vehicles, plus many more. The Norwegian Navy took part in the exercise with the following ships: two fast frigates named "HNOMS Bergen" and "Narvik", the mine countermeasure ships "HNOMS Vidar", "HNOMS Alta", "HNOMS Oksoey", "HNOMS Karmoey" and "HNOMS Otra", the submarine "HNOMS Utsira", three CB-90 combat boats, and fast patrol boats of 22 MTB Squadron, as well as one further unnamed submarine. The Norwegian Air Force was involved in the exercise with two DA 20

In order to support amphibious landing operations, the RNLMC can field two Beach Armored Recovery Vehicles (BARVs). Both Dutch BARVs are permanently based on the Landing Platform Dock "HNLMS Rotterdam". The BARV is used as a recovery vehicle in the surf zone and on the beach. The vehicle is able to push grounded landing craft back into deeper water. It can also recover vehicles that have been stranded after leaving a landing craft. The BARV is based on the former Dutch Army Leopard 1V main battle tank chassis, which has been waterproofed. Modifications to the BARV include a steel crew cabin with bullet-proof glass, an outside working deck, modified running gear and new tracks, as well as cooling and exhaust stacks to allow deepwater operations. The pictured vehicle bears the nickname "Hercules", while the second Dutch BARV is nicknamed "Samson".

In the dock of "RNLMS Rotterdam", the ship carries four Landing Craft Utilities (LCUs). Here, one of these LCUs can be seen ferrying vehicles ashore during the NEO operation.

During the NEO, Norwegian CB-90H combat boats patrolled the sea and transported the civilians from the shore to "HNLMS Rotterdam". Developed by the Swedish shipbuilder, Dockstavarvet, CB-90s were designed as a fast attack boat, patrol boat and special operations support vessel. The CB-90H boat uses two water-jet propulsion units, each powered by a 460kW diesel engine, developing sustained speeds in excess of 40 knots and with high maneuverability. Inside the crew compartment of the CB-90H combat boat, up to twenty-one fully equipped infantrymen, or 4.5 tons of cargo can be transported. While other weapon systems can be mounted, the standard armament of the CB-90H consists of a 12.7mm M2HB machine gun.

electronic warfare aircraft, sixteen F-16 fighter aircraft, two S61 SAR helicopters, eight Bell 412 transport helicopters, one Lynx naval helicopter, one P-3C Orion anti-submarine warfare aircraft, and two C-130 Hercules transport aircraft. In addition, the Air Force took part in the exercise with its NRF-earmarked ground-based air defense formation with the NASAMS air-defense system. Civilian volunteers and members of the Norwegian Red Cross took part in the exercise as role players, and acted as Trondian or Ranian civilians, or foreigners which had to be evacuated by the MNTF. Overall command and control of the MNTF was provided by the Norwegian Joint Operations Headquarters in Stavanger.

A challenging scenario

The scenario for Exercise Battle Griffin 2005 was based on crises such as the war in Afghanistan and the associated ISAF operation, the Kosovo war and the implementation of KFOR, or the U.S.-led war against terrorism in Iraq. For the scenario, Norway was divided into the countries of Northland, Utopia and Southland. The three generic countries split along ethnic lines several years ago in 1987. The breakup of the three countries was peaceful. However, since becoming an independent country, Utopia developed territorial desires for the northern part of Southland. These claims were based on the fact that there were ethnic bonds between the inhabitants of southern Utopia and the people living in the northern part of Southland. Simultaneously, in the multiethnic southern province of Utopia (known as Trondia), uneasiness between the ethnic Trondians and ethnic Ranians residents suddenly erupted. This led to instability and unrest in the region. The spreading of conflict and unrest within Utopia became a major concern for Southland. The discovery of oil resources near the coast of the province of Trondia in 1991 further complicated the situation. By November 2004 there was tension between the provincial government of Trondia and the Utopian government about the use of the oil resources. In its first two United Nations Security Council Resolutions - namely UNSCR 6000 and UNSCR 6001 - the international community expressed its concerns and called for a peaceful solution in order to bring stability back to the troubled region. But the situation was getting worse and by January 2005, Resolution 6002 was approved by the United Nations Security Council. In this an embargo was authorized. In the following weeks the ethnic unrest reached a new level, with extremist Trondian groups using harsh means to threaten Ranian minorities in Trondia. In return, entire Trondian villages were destroyed in Rania. The Utopian armed forces deployed into Trondia so that they could protect Ranian villages and enclaves. The region was close to the breakout of civil war, which might also have spread into Southland.

Main armament of the Norwegian CB-90 combat boat is the 12.7mm M2HB machine gun. In addition, the boat can be fitted with a variety of weapons such as the 40mm automatic grenade launcher manufactured by Heckler and Koch.

Due to its construction, the CB-90 combat boat is able to land at nearly any point of the coastline. Here a CB-90 can be seen closing in on a small cliff. The front troop hatch is already open, through which up to twenty-one fully equipped infantrymen can board the vessel. The Norwegian Armed Forces use their CB-90 combat boats as fast attack boats, patrol boats, and special operations support vessels.

At high speed, the rigid inflatable boats of the hostage rescue force close in on the hijacked vessel. The operatives on board the rigid inflatable boats, provided by the Dutch Bijzondere Bijstands Eenheid (BBE), can be seen busily preparing for the boarding procedure. The rigid inflatable boat in the rear belongs to the Dutch Marines, whereas the one in front is of Norwegian origin.

The Norwegian rigid inflatable boat with its Dutch operatives has reached the ship. The operatives are just in the process of hooking a rope ladder to the side of the hijacked ship. This photo shows the critical part of the boarding operation, in which the hostage rescue force is at its most vulnerable.

Mission of the Multinational Task Force

According to the exercise scenario, the UN Security Council authorized a Multinational Task Force (MNTF) to lead and conduct military operations under Chapter VII of the UN Charter in cooperation with the governments of Utopia and Southland. This course of action was laid out in Resolution 6003, enacted in February 2005. The mission of the MNTF was to separate the parties in Southern Utopia, put an end to the ethnic cleansing, and maintain peace and security in the region. Command of the MNTF was given to Norway and a total of fourteen nations contributed troops to the joint force that incorporated ground, naval and air assets.

Phases of Exercise Battle Griffin 2005

Beginning on 21 February 2005, Exercise Battle Griffin 2005 consisted of four phases. The main objectives of phase one, the **Deployment Phase**, were as follows:

- Deployment of all forces of the MNTF into Southland.
- Conduct essential information operations in order to gather information on the situation in the troubled region, and to inform the local population about the presence of the MNTF and its mission.
- Establish and enforce the UNSCR embargo.
- Establish the necessary control over the sea and air space in the troubled region in order to enforce the previously established embargo, and to prepare the way for a safe forward deployment of the MNTF.
- Present a credible show of force in order to deter hardliners from conducting further acts of aggression.
- Conduct Joint CET/FIT

This phase ended when all forces were in position, and sea and air superiority was established. During this phase, all units had begun with interoperability training so as to prepare for their movements into the troubled region of Utopia. On 2 March 2005, phase two of Exercise Battle Griffin 2005, the **Stabilization Phase**, started. This phase included the following tasks:

- Maintaining the embargo by the use of naval and air force assets of the joint combined MNTF. At the same time, maintain control of the air and sea space in order to secure the deployment of ground forces.
- Conduct initial entry operations and amphibious operations in order to support the deployment of ground forces into the troubled region of Utopia.
- Movement of ground forces of the MNTF into Trondia, the troubled region of Utopia, in order to establish a safe and secure environment for the local population.
- Conduct stabilization operations in hotspots of the troubled region. This included twenty-eight enclaves sheltering different ethnic groups. Trondian enclaves in mostly Ranian populated areas were threatened, while Ranian enclaves in mostly Trondian populated areas were also being threatened. Other hotspots included areas that

In a Joint/Combined Special Forces operation, a hostage rescue operation was conducted on a ship. The ship of the MNTF had been boarded by a group of terrorists and its crew taken hostage. During the operation, operatives of the Dutch BBE were deployed simultaneously to the hijacked ship from a Swedish helicopter and Norwegian rigid inflatable boats.

This close-up of a Norwegian Special Forces rigid inflatable boat illustrates well the firepower of the craft. In the front of the boat, a 12.7mm M2HB heavy machine gun is mounted. In the rear, a 40mm automatic grenade launcher can be seen.

This crewmember of a Norwegian CB-90 combat boat is armed with a 5.56mm G36kA1 carbine version of the G36 assault rifle. The weapon is fitted with an Aimpoint CompMXD reflex sight.

hosted hardliners and militias with unknown aims and political objectives. And finally, there were areas that mainly supported the Trondian Protection Corps (TPC), a local militia.

- Conduct necessary Counter-Terrorism Operations.
- Conduct necessary operations against Utopian forces in order to ensure freedom of movement for the MNTF and Non-Governmental Organizations.

This second phase ended on 8 March 2005, and included all the types of missions and tasks which would normally appear in a real Crisis Response Operation (CRO) or Peace Support Operation (PSO). At the beginning of the phase, the troops had to negotiate their way through the region in order to reach their Areas of Responsibility (AORs). Here they were faced by local militias (played by members of the Norwegian Home Guard), which in some cases blocked deployment roads in order to extort a "road toll", or to gain any type of support they could, such as weapons, food, medical care, or the granting of rights for power in the region. The exercise troops also encountered minefields, unexploded ordnance on roadsides, damaged bridges and anti-UN demonstrations. Another factor complicating the situation was the presence of Utopian armored forces in the area, and several times, agreements had to be reached with them to prevent point blank standoff situations.

Once the troops were established in their AORs, they had to stop the ethnically motivated violence. Again this was not easy, as the exercise controllers had prepared a number of incidents. Ethnic enclaves were attacked and burnt down. Militias conducted illegal checkpoints. Violent demonstrations were conducted to protest against the presence of the MNTF. Attacks on the MNTF were conducted by hardliners, including vehicle-borne improvised explosive device attacks (VBIEDs), drive-by shootings, or improvised explosive devices (IEDs) placed at the side of roads. The MNTF had to secure the deployment and distribution of aid into the region. However, according to the exercise scenario, the MNTF failed to stabilize the situation in the first place, and tension in the region quickly escalated.

Local militia protecting a Ranian enclave have blocked a road of the MNTF and stopped a convoy. Now it is the convoy leader's job to negotiate a way through for his unit. This is not an easy task, as an earlier convoy commander had made some promises which had not yet been fulfilled. The leader of the militia demands protection from the MNTF and humanitarian help including food. The Norwegians often did an outstanding job in providing the exercise troops with realistic incidents.

An Italian mountain infantry company of the 9th Alpini Regiment from L'Aouila also took part in the exercise. For transportation, the Italian company used Bv 206 all-terrain vehicles supplied by the Norwegian Army. Here, some of these Bv 206s can be seen during a patrol in a small village. The machine gunner on the vehicle is armed with a 7.62mm MG42/59 machine gun.

These troops belong to the 93rd Company of the 9th Alpini Regiment from L'Aouila. The Italian Alpini are experts in mountain and arctic warfare, this being well reflected in their equipment that includes snow camouflage suits, combat vests, and cold weather protective clothing. The Alpini in the foreground is armed with a 5.56mm Beretta 70/90 assault rifle.

Another weapon used by the Italian Alpini mountain infantry is the 5.56mm MINIMI light machine gun.

Due to the rise in violence, it was decided by the MNTF to conduct a Non-combatant Evacuation Operation (NEO) in order to extract foreign nationals from the region. The NEO took place on 5 March, and an amphibious operation was conducted by the Dutch Marines, who were acting as a reserve for the MNTF. Finally, agreements with Utopian forces for their withdrawal from the region had to be made. After the rise of violence in the region, the Utopian commander claimed that the MNTF was unable to provide security for the resident Utopian population. Therefore, he refused to remove his troops and threatened to conduct aggressive operations in the region in order to re-establish law and order himself. Obviously this was a course of action that totally opposed the UN plan and resolution. Due to this situation of regional escalation, the MNTF had to conduct a full-scale combat operation to show its strength, and by this, force the withdrawal of Utopian units. In clashes between Utopia and the MNTF, heavy armor was used, and full-scale war fighting skills were required from the MNTF forces. The phase ended on 8 March 2005 with the partial neutralization of the Utopian armed forces positioned in Trondia. The phase resulted in an end to ethnic cleansing in the region, and the MNTF was left in charge of the unstable "hotspots". Finally, the different factions in the region complied with the MNTF.

Phase three, also named the **Implementation Phase**, saw the MNTF changing its task from full-scale war operations back to peace support operations. During this, the following tasks had to be fulfilled:

- Maintain the embargo and the control of air and sea space in order to provide security for the forces operating on the ground.
- Maintain and improve a stable and secure environment for the local population of Utopia.
- Prepare for a handover to the UN and local governments.
- Liaison with Governmental and Non-Governmental Organizations in order to allow a quick rebuilding of the region.

Conducted on 8 - 9 March, this phase basically saw troops improving the situations in their AORs. During this third phase, troops mainly conducted security patrols, maintained contacts with local leaders, and supported preparations for the re-establishment of local governments. All kinds of humanitarian support were also provided, such as the distribution of food, the building of bridges by engineers, and so on. This phase ended on 9 March, when it was stated that UNSCR 6003 had been fully implemented, and that local governments were now prepared to provide security within the troubled region of Utopia. It must be said that here the scenario was rushed a little, if comparing it with reality, as we all know from Bosnia and Kosovo that such a process might take years. The final or fourth phase of Exercise Battle Griffin 2005 was the **Transition and Exit Phase**. This included the redeployment of the different forces of the MNTF back to their home countries.

Special Forces operations

Special Forces assets from Norway (Army, Navy and 137th Air Wing with helicopters), the USA (SEALs), Finland (Army and helicopters), the Netherlands (Special Assistance Unit), and Sweden (Army and helicopters) were also taking part in Exercise Battle Griffin 2005. In total, some 520 Special Forces personnel were involved in the exercise. Kept well out of the sight of the media, the multinational Special Forces Contingent supported the MNTF during its operations. Based some hundred kilometers to the southwest of the exercise, at the Norwegian Air Force Base at Orland, the Special Forces trained in long-range insertions before conducting their missions. While most of the training of the Special Forces was kept secret, some parts were revealed to the public. For example, during the exercise, Special

The Swedish mechanized infantry company of I5 Regiment "Fältjägarna", based at Östersund, deployed to Exercise Battle Griffin 2005 with ten CV 9040 infantry fighting vehicles. The CV 9040 is armed with a 40mm Bofors L/70 cannon, and a coaxial 7.62mm machine gun. Here, two CV 9040s of the unit can be seen providing security while an arrest operation is under way.

Forces were used for hostage rescue operations on land and on sea. In one specific case, the Dutch Special Assistance Unit was tasked with boarding a ship that had been seized by terrorists. The troops stormed the ship with the aid of speedboats and helicopters. The boats were from Norwegian Special Forces assets, while the helicopters were Swedish. The operation also showed another aim of the Special Forces training, that of Joint Combined training. Also conducted by the Special Forces during the exercise was the gathering of key information on terrorist activities and the situation in the troubled region, the seizing of oil platforms from irregular forces in the disputed area, and the arrest of key anti-MNTF activists.

ISTAR Battalion

In Exercise Battle Griffin 2005, Norway for the first time tested its new Intelligence, Surveillance, Target Acquisition and Reconnaissance (ISTAR) Battalion. The unit was designed to supply intelligence and sensory data to land, naval and air forces, helping to keep them all on the same tactical page. Basically the mission is simple, gather information about the enemy, create a picture that is as accurate as possible of the enemy forces and their intents, and then make it accessible to their own forces. The beneficiary forces of this data could be either Norwegian or NATO units. Within the Norwegian Army, the ISTAR concept is something totally new. The ISTAR Battalion is formed from elements of the Norwegian Rangers, FAC (Forward Air Controller) teams, and an electronic warfare component. A UAV element with Unmanned Aerial Vehicles will also be part of the ISTAR Battalion in the near future. In fact, the ISTAR Battalion basically works as a "toolbox", making sure that all the components are working together. The mission of the Battalion's command post is to analyze the information that has been gathered, and form it into a single picture. Under development for two years, the ISTAR Battalion collects targeting information and assesses battle damage. It scoops up data via a range of means that are at its disposal, including long-range patrols and other ground reconnaissance, the Arthur artillery-and-small-arms locating radar and other types of radar, electronic warfare, F-16s and P-3 Orions, electronic warfare systems, and satellites. The Battalion can also be supported by NATO airborne warning and control systems. Under the Battalion's headquarters, among other elements, are situated a long-range reconnaissance company and an electronic warfare company.

Marine Air Ground Task Force 25

The U.S. contribution to Exercise Battle Griffin 2005 was Marine Air Ground Task Force 25. MAGTF-25, as it is known in its abbreviated form, is part of the 4th Marine Division based at New Orleans. The division is composed entirely of units of the Marine Reserve. The force package of

In this photo, a Swedish infantryman can be seen in the open hatch of the rear compartment of a CV 9040 infantry fighting vehicle. He is armed with a 5.56mm AK5 assault rifle. The AK5 assault rifle is based on the FNC assault rifle and is manufactured by Bofors Carl Gustaf AB.

In addition to the infantry fighting vehicle version of the CV 90, the Swedish Army uses the vehicle in a couple of specialized versions. One of these is the CV 90 Armored Recovery Vehicle (ARV). The ARV is fitted with a front-mounted dozer blade, and two hydraulic winches.

Also deployed to Norway by the Swedish troops was this MT-LB, known by the Swedes as Pbv 401. Sweden received over 1,000 MT-LBs from the former East German Army. While half of the vehicles were scrapped for spare parts, the remainder was converted into command vehicles, logistic support vehicles and ambulances.

Another vehicle used by I5 Regiment "Fältjägarna" during Exercise Battle Griffin 2005 was this MT-LBU ambulance. It is believed that only twenty-five vehicles of this type are in service with the Swedish Armed Forces. The MT-LBU ambulances were converted out of MT-LBU command post variants that Sweden received from former East German Army stocks.

MAGTF-25 consisted of a Command Element (150 troops), a Ground Combat Element (700 troops), an Aviation Combat Element (150 troops), and a Combat Service Support Element (300 troops). The Command Element and the Ground Combat Element of MAGTF-25 in Exercise Battle Griffin 2005 were provided by a battalion from the 25th Marine Regiment, based at Worchester, Massachusetts. The force was supported by a company-sized formation of the 4th Light Armored Reconnaissance Battalion with nine LAVs. While the U.S. Marines deployed their own light armored vehicles, HMMWVs and heavy trucks, they were also equipped during the exercise with additional Bv 206 all-terrain vehicles. The Bv 206s were provided by the Norwegian Armed Forces, together with a complement of Norwegian drivers. The Air Combat Element of MAGTF-25 was formed of elements of Marine Aircraft Group 41, namely six F/A-18 fighter/attack aircraft of VMFA112 (Marine Fighter Attack Squadron 112 "Cowboys"), and two KC-130 aerial refueling aircraft of VMGR 234 (Marine Aerial Refueler Transport Squadron 234 "Ranger"). In peacetime, both units are based at the Naval Air Station-Joint Reserve Base Fort Worth. Throughout this exercise, the Air Combat Element of MAGTF-25 was partly based at the Norwegian airfields at Orland and Gardermoen. The Combat Service Support Element was provided by elements of different units of the 4th Force Service Support Group. For the Marines of MAGTF-25, Exercise Battle Griffin 2005 provided a welcome opportunity to hone their edge in the art of arctic warfare and peace support operations in arctic environments. In fact, the 25th Marine Regiment is one of the units of the United States Marine Corps Reserve that is earmarked for operations in arctic conditions. In addition, the exercise was used to familiarize new Marines with the regiment, and to train troops at the unit level. Since the beginning of "Operation Iraqi Freedom" and the deployment of parts of the 4th Marine Division to Iraq, training at the unit level has become a rare sight, and therefore their deployment to Norway was greatly appreciated by the Marines and their

This Patria XA-180 6x6 wheeled armored personnel carrier, belonging to the Finnish Rapid Deployment Force, was seen guarding the Headquarters of the Finnish-Swedish Battalion. The XA-180 has a combat weight of 22 tons and is powered by a Valmet six-cylinder turbo diesel engine. The vehicle can reach a top road speed of 90km/h and has a road range of 800km. The XA-180 is fully amphibious, and when afloat it is powered by two propellers.

The XA-180 6x6 wheeled armored personnel carrier is widely used by this Scandinavian army as the vehicle of choice for peace support operations. Therefore, it is no wonder that the FRDF deployed with XA-180s on Exercise Battle Griffin 2005. Here, an XA-180 and its crew can be seen guarding a crossroads in Namsos. In addition to a three-man crew, the vehicle can transport eight fully equipped infantrymen. Thanks to its V-shaped hull, the vehicle has a decent degree of protection against the blast effects of mines. Its hull protects the crew against small-arms fire and the effects of artillery splinters.

On the roof of the XA-180 wheeled armored personnel carrier, a 12.7mm NSV heavy machine gun is mounted. The NSV uses a 12.7x107mm cartridge, and the weapon has an effective range of up to 2,000m. The weapon's cyclic rate of fire is 700 to 800 rounds per minute, and the NSV has a muzzle velocity of 820-860m/s.

officers. During an interview, the commander of MAGTF-25, Colonel Joseph L. Osterman, outlined his aims for the exercise as follows: *"My intent is to maximize our training opportunity in a joint and combined environment, focusing on MAGTF doctrine and the full spectrum of war. We must take every opportunity to capitalize on the multinational dimension of the exercise to improve interoperability. The end state desired is an MAGTF capable of efficient NATO operations, Marines prepared for Security and Stabilization Operations (SASO) through offensive operations, and multinational partners conversant in unique attributes of MAGTF doctrine."*

Opposing forces

Opposing the MNTF during Exercise Battle Griffin 2005 were the forces of Utopia. It was these troops that confronted the MNTF armored units in full-scale war-type clashes on 8 March 2005. The force consisted of all the German Air Force units involved in the exercise. Also, part of the Utopian force was Combined Air Operations Center 3 (CAOC3) of NATO, which acted as the Utopian air force operations center. Other air force assets acting as Utopian role players included four Norwegian F-16 fighter aircraft, eight Swedish JAS Saab Gripen fighter aircraft, and two S61 SAR helicopters. From RAF Mildenhall in the UK, two KC-135 tankers of the USAF flew air refueling missions for the Utopian air force. On the navy side, the Utopian forces consisted of the fast patrol boats of 22 MTB Squadron, as well as one unnamed Norwegian submarine. On the ground, the Utopian army units were played by a Norwegian mechanized infantry battalion and units of the Norwegian Home Guard.

Finnish Swedish Task Force

The 6th Norwegian Division, which acted as headquarters of the land forces of the MNTF, had three battalion-sized task forces under its command. One was formed from a Norwegian mechanized infantry battalion, augmented by an Italian mountain infantry company. The second task force was formed by MAGTF-25 of the U.S. Marine Corps. The third one was a battalion-sized formation built from units of the Finnish Rapid Deployment Force and a mechanized infantry company from Sweden. The Swedish company was drawn from I5 Regiment "Fältjägarna", based at Östersund. Together with some staff officers and a reconnaissance platoon (part of the HQ company), the Swedish troops in the bi-national battalion numbered 178 troops. Heavy equipment of the Swedish troops consisted of CV 9040 infantry fighting vehicles, one CV 90 recovery vehicle, one MT-LB and one MT-LBU. Supporting the Swedish company during the exercise was a Finnish signals detachment. The Finnish portion of the bi-national battalion consisted of a battalion headquarters and support company, and a mechanized infantry company. In total, 400 Finnish soldiers with seventy-eight wheeled vehicles contributed to the battalion. Twenty of the wheeled vehicles were Patria wheeled armored personnel carriers. Within the headquarters and support company were situated a mortar platoon, a reconnaissance platoon, an EOD team and a signals platoon. The Finnish mechanized infantry company was augmented by a military police contingent that had to utilize its skills when it came to contacts with war criminals and insurgents.

This well-camouflaged soldier belongs to the reconnaissance platoon of the Finnish Rapid Deployment Force. The soldiers of this platoon provided the eyes and ears of the Finnish-Swedish battalion, and they conducted a variety of missions, including the observation of locations in which war criminals were located. The soldier is armed with a 5.56mm M95 SAKO assault rifle.

Troops of the MNTF during Exercise Battle Griffin 2005 were kept busy protecting the various ethnic enclaves. This Finnish soldier of the FRDF is armed with a 5.56mm light machine gun.

Helicopter Task Group

Several nations deployed helicopters and ground crews to Norway as part of this major exercise. On the coalition force's side, these were formed into a multinational Helicopter Task Group (HTG) based at Orland. The mission of the HTG was to support the units of the MNTF with air transport capability, reconnaissance gathering, and liaison operations. This included the insertion and extraction of Special Forces, as well as the transportation of quick reaction forces during crisis situations. In addition to operational challenges during Exercise Battle Griffin 2005, the helicopter crews also had to deal with environmental challenges such as snowstorms and temperatures well below zero. The HTG was formed under the command of a small Norwegian HQ. It had command of the following assets:

- Finland: two Mi-8 transport helicopters and four MD-500 liaison helicopters.
- Sweden: two CH-46 transport helicopters and two BO105 liaison helicopters.
- Germany: two Bell UH-1D light transport helicopters
- Denmark: four Fennec combat helicopters
- Norway: eight Bell 412ST light transport helicopters

This total of twenty-four helicopters was supported by 199 troops, a total which included both the aircrews and the ground personnel.

Non-combatant Evacuation Operation (NEO)

After tensions in the troubled region had risen to an unacceptable level, the MNTF decided to evacuate all foreign non-combatants from Trondia. Tasked with this mission was the amphibious component of the MNTF, provided by elements of the Royal Netherlands Marine Corps. After the civilians of the various nations were registered by their ambassadors, they were gathered together, and on 5 March, transported by buses to the extraction point. In the meantime, early in the morning of the same day, the Dutch Marines had landed in a remote area 60km west of Namsos in order to establish a security perimeter around a small natural harbor which would soon become the key to the Non-combatant Evacuation Operation. The Marines set up three security rings, an outer one protecting the only bridge leading to the island on which the harbor was situated, an extended one which covered all approaches to the harbor with heavy infantry weapons, and a third one that used dismounted troops to provide a close protection ring for the NEO site. At sea, the area was patrolled by CB-90 combat boats and other vessels. The air space over the NEO site was also under constant protection from fighter aircraft of the aviation element of the MNTF.

Close to the harbor that was serving as an extraction point, a couple of tents were erected in which a screening and registration center was setup. By nine o'clock in the morning, buses with nearly 140 civilians aboard arrived, and troops began the evacuation process. Before being shipped by landing craft or combat boats to the Landing Platform Dock "HNLMS Rotterdam", which was acting as a safe haven, the civilians underwent registration as well as a security and identity check in the screening center. In addition, medical assistance was provided to injured people. During this part of the NEO, the participating troops were faced with several incidents organized by Exercise Control (EXCON). For example, several of the civilians were wounded and had to be carried on stretchers, while some of the people did not possess any identification papers, and others resisted the evacuation for several reasons. In another incident, the Dutch Marines had to deal with a terrorist who had managed to infiltrate amongst the civilians. The terrorist, a female, was equipped with a bomb strapped

From the commander's cupola of an XA-180 wheeled armored personnel carrier, a Finnish platoon commander uses binoculars to observe a group of potential troublemakers. Note the Rapid Deployment Force shoulder title, Finnish national flag, and the formation badge worn on his left upper sleeve.

In Namsos, troops of the Finnish Rapid Deployment Force guard an important bridge. With countless checkpoints, the MNTF established control over Trondian territory. At the same time the checkpoints were used to monitor the progress of the Utopian force withdrawal from the region. Another task of the checkpoints was to deny Utopian forces any opportunity of returning to any previously cleared regions. In some cases this last task led to standoff situations between the commanders of Utopian forces and the MNTF.

to her body. This was discovered during the initial security checks and an EOD team was called to deal with the bomb. While the civilians were checked through the registration camps, the security perimeter of the Marines became the focus for another incident. Here, a mob of people equipped with sticks and stones and yelling anti-MNTF slogans, gathered and started to get more and more aggressive towards the guards of the Dutch Marine Corps. While reinforcements were still on their way, the mob attacked the Marines blocking the road to the extraction point and the registration center. The Marines first used non-lethal means to disperse the crowd, but when this did not work, they fired warning shots. This stopped the progress of the mob, and when reinforcements arrived, the superiority of force deterred the aggressors from any further action.

In the meantime, the first civilians were shipped by LCUs (Landing Craft Utilities) and CB-90 combat boats to "HNLMS Rotterdam". On board the Landing Platform Dock, sailors awaited the civilians, who were then required to register again and undergo a security check. Following this process, the civilians were quartered on board, received a hot meal, and were briefed on their journey out of the crisis region. The wounded civilians were transferred to the ship's hospital where medical treatment was given to them. During the exercise, the ship's hospital was reinforced by a Norwegian surgical team. Of interest was the mix of civilians supplied by the Norwegian Red Cross for the exercise. The number of young and old people with different levels of English language knowledge provided a real challenge for the exercise troops, in some cases. On the ship, one nurse stated to the author: "It is good training for us, especially because in some cases we can gather experience in how to deal with foreign people who are not accustomed to our language, military habits and might not have been on a naval ship before." In an interview, a Dutch public affairs officer summarized the reasons for the participation of the Netherlands in Exercise Battle Griffin 2005:

"*For the Netherlands, it is important to train with the Norwegian Armed Forces. Within NATO it is very important that procedures are tuned and that we train together. Combined Joint exercises are important for real life operations. The biggest challenge during the exercise for the Netherlands participants would be to operate in our new organization as the Netherlands Maritime Force (NLMARFOR) for the first time, with an almost fully Dutch CATF/CLF (Commander Amphibious Task Force/Commander Landing Force) staff. It is a challenge to train ourselves in this new situation at all levels and in all facets of amphibious operations. Our main goal is to become an integrated and coordinated maritime expeditionary unit.*"

Another picture of a Patria XA-180 6x6 wheeled armored personnel carrier of the Finnish Rapid Deployment Force. The fully amphibious XA-180 has a combat weight of 22 tons and is powered by a Valmet six-cylinder turbo diesel engine. The vehicle can reach a top road speed of 90km/h and has a road range of 800km. On the roof of the XA-180 wheeled armored personnel carrier, a 12.7mm NSV heavy machine gun is mounted.

Land Rover Defender 110 vehicles were used by the Finnish Rapid Deployment Force as liaison vehicles and as cross-country transport. Interestingly, the vehicles were painted in a glossy, light green color.

Communications within the Finnish-Swedish Battalion were provided by a Finnish signals platoon. The signals platoon transported their radio equipment on SISU NA-110 all-terrain vehicles. The SISU NA-110 is very similar to the Hägglunds Bv 206. Consisting of a front and rear unit, the vehicle can transport up to eighteen troops, or 2.2 tons of equipment. Powering the NA-110 is a V-8 diesel engine that develops 150hp, and provides the vehicle with a top speed of 55km/h. The NA-110 is fully amphibious.

The deployment of the MNTF into Trondia was spearheaded by reconnaissance assets of the armor battalion of Brigade North. Here, Mercedes GD240 4x4 cross-country vehicles of a light reconnaissance platoon can be seen. Even with temperatures well below zero, the reconnaissance troops are riding in their vehicle with the windshield folded down and the soft cover partly removed. Note the wire guards which protect the crews against stones thrown at them; they were introduced for peace support operations after incidents in Bosnia and Kosovo caused several injuries.

Another picture showing Mercedes GD240 4x4 cross-country vehicles of a light reconnaissance platoon. The vehicles are armed with two 7.62mm MG3 machine guns, one of which is mounted in front of the passenger seat. The vehicle crews consist of a driver, a commander and a machine gunner. In the rear of the vehicles, light anti-tank weapons are carried. Note the smoke-laying system and the infrared lights for driving with night vision equipment.

Here a vehicle checkpoint conducted by a light reconnaissance platoon of the armor battalion of Brigade North can be seen in action. While the machine gunner on the Mercedes GD 4x4 cross-country vehicle provides cover, the rest of the crews can be seen conducting ID checks and vehicle searches. During Exercise Battle Griffin 2005, troops of the MNTF conducted countless vehicle checkpoints in the exercise area. Most of these were placed on public roads and in some cases even local non-exercise civilian traffic was checked. But the efforts of the MNTF paid off, since some wanted suspects were arrested, weapons were seized, and smuggling was made much more difficult.

In an exercise break, a tanker from the 1st Escadron "Panser Battaljonen i Nord Norge" enjoys a hot meal. The battalion has two companies equipped with Leopard 1A5NO2s. Within the companies, the soldiers are called dragoons. Note the blue leopard's head and turret number painted on the vehicle.

Now partly replaced by Leopard 2A4 main battle tanks, the Leopard 1 in various guises has been the Norwegian main battle tank since 1971. The latest Leopard 1 model in service with the Norwegian Army is the Leopard 1A5NO2, which can be seen here. The red crosses identify the vehicle as one attached to the Utopian armed forces that provided the opposing force for the MNTF.

A Leopard 1A5NO2 of the "Panser Battaljonen i Nord Norge" during Exercise Battle Griffin 2005. The vehicle is partly covered in a winter camouflage pattern. For winter camouflage the Norwegian Army uses a whitewash which can be easily applied and removed.

A rear view of a Leopard 1A5NO2 - clearly visible is the add-on armor applied to the vehicle's turret. The vehicle belongs to the Badufoss-based "Panser Battaljonen i Nord Norge", which during Exercise Battle Griffin 2005, acted as Utopian armed forces.

Even with the more modern Leopard 2A4 recently entering Norwegian Army inventories, it is believed that the Leopard 1A5NO2 will stay on in Norwegian service for several more years. With a much lower combat weight, the Leopard 1A5NO2 is much better suited to the small bridges which can be found all over the difficult terrain of central and northern Norway. The Leopard 1A5NO2 has a similar fire control system to the Leopard 2A4, but it is outgunned by the new tank's 120mm smoothbore cannon. In terms of speed and agility, the Leopard 1A5NO2 is more then equal to the Leopard 2A4. In this picture, some of the capabilities of the Leopard 1A5NO can be observed.

A column of Leopard 1A5NOs moves down a road in central Norway. The vehicles are acting as Utopian forces and have the mission of challenging MNTF units, in order to provoke the exercise troops into some type of hostile action. While in the final phase of Battle Griffin 2005, the MNTF had to engage the Utopian forces in order to force their withdrawal. In the early parts of the deployment of the peace force, the MNTF unit commanders had to negotiate with Utopian forces in order to prevent any clashes.

For recovery and maintenance work within the Norwegian armored units, the Bergepanzer 2 Armored Recovery Vehicle (ARV) is used. Amongst others, the Bergepanzer 2 can fulfill the following tasks: recover vehicles, refuel and de-fuel vehicles, lift equipment up to a weight of 20 tons and up to seven meters above the ground, change power packs, lift and change turrets, remove roadblocks, dig field positions for tanks, and tow disabled vehicles. On-board equipment of the ARV includes welding and cutting gear, a fuel pump, hydraulically operated dozer blade, a crane, and a winch.

Also deployed in Exercise Battle Griffin 2005 were a couple of Leguan AVLBs (Armored Vehicle-Launched Bridges). Developed from the Leopard 1-based "Biber" bridge-layer, the Leguan bridge can span 26 meters and be used by vehicles up to MLC 70. In total, nine tracked Leguan AVLBs are in service with the Norwegian Army. The pictured vehicle belongs to an armored engineer unit of the 6th Norwegian Division. With its bridge, the AVLB has a combat weight of nearly 50 tons. Its crew consists of a driver and a vehicle commander. Laying the bridge can be performed in less then five minutes. During the exercise, the Leguans were often used by the MNTF to cross destroyed bridges in order to speed up the advance of the peace force during its deployment to the crisis regions.

In addition to the tracked version of the Leguan bridge, the Norwegian Army also fields a wheeled version of the bridging system. The wheeled Leguan uses the same bridge as its tracked brother, with a 26m span and a load capacity of MLC70. The wheeled Leguan is based on a MAN VFAE 8x8 chassis, powered by a 412hp diesel engine. The vehicle has a combat weight of nearly 36 tons and can travel at a maximum speed of 70km/h. Together with a road range of 600km, this makes the wheeled version of the Leguan much more suitable for low intensity conflict scenarios were bridges have to be built, but where there is no enemy threat at the bridging site.

Another engineer vehicle of the Norwegian Army, which is based on the Leopard 1 chassis, is the Armored Engineer Vehicle (AEV). A total of twenty-two AEVs are in service with the Norwegian Army. The vehicles were converted by Hägglunds from Leopard 1A1NO main battle tanks. Main features of the Leopard 1 AEV are a centrally mounted excavator arm with a 1m_ bucket. Other equipment of the AEV includes a 3.6m wide dozer blade, welding/cutting equipment, and two hydraulic winches. The AEV crew consists of a driver and a vehicle commander. Basic tasks of the AEV include obstacle clearing, construction of anti-tank ditches, preparation of bridging sites for AVLBs, and improving roads and routes.

This Narvik Technology AS P6 all-terrain vehicle belongs to the inventory of the Norwegian ISTAR Battalion. Similar to the Bv 206, the vehicle consists of a front and rear cab. In the rear cab, an electronic warfare device is situated. The device is a frequency jammer that can be used to disrupt enemy communications by jamming their frequencies.

Like the Army, the USMC uses the M998 HMMWV with a four-door soft-top enclosure as its standard cross-country vehicle. The HMMWV vehicles of the USMC can be identified by their extended exhaust pipes, items which can be clearly seen here.

Just before his battalion moves out on a mission, the battalion commander of MAGTF-25's ground component conducts a final officers' briefing.

The anti-tank platoon of the infantry battalion of MAGTF-25 is equipped with M1046A1 HMMWV TOW carriers with supplemental armor and a winch. With their TOW anti-tank missile systems, the Marines can destroy any known armor out to a distance of 3,750m. The thermal imaging sight of the TOW system is also a very effective surveillance aid.

Also deployed to Norway by D Company of the USMC's 4th Light Armored Reconnaissance Battalion was one LAV-M mortar carrier. Instead of a turret, the vehicle has a large hatch in the center of the hull roof and an 81mm mortar is mounted inside on the vehicle floor. On board, 94 rounds for the mortar are carried. In addition, the vehicle is armed with a pintle-mounted M240 machine gun and two M257 four-barreled smoke grenade launchers.

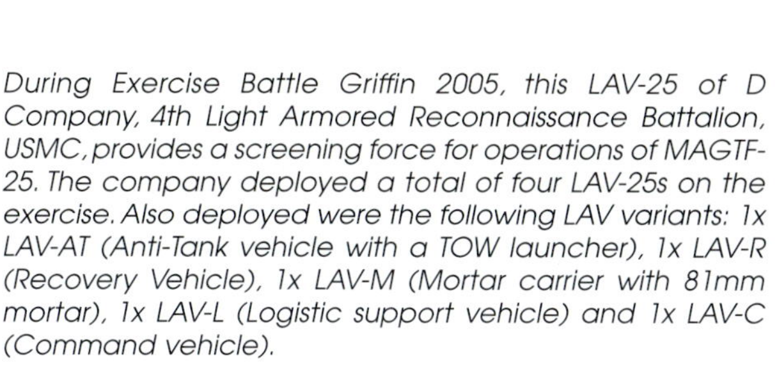

During Exercise Battle Griffin 2005, this LAV-25 of D Company, 4th Light Armored Reconnaissance Battalion, USMC, provides a screening force for operations of MAGTF-25. The company deployed a total of four LAV-25s on the exercise. Also deployed were the following LAV variants: 1x LAV-AT (Anti-Tank vehicle with a TOW launcher), 1x LAV-R (Recovery Vehicle), 1x LAV-M (Mortar carrier with 81mm mortar), 1x LAV-L (Logistic support vehicle) and 1x LAV-C (Command vehicle).

The LAV-25 is fitted with a two-man turret and armed with a 25mm M242 chain gun, and a 7.62mm M240 coaxial machine gun. The vehicle has a combat weight of 13 tons and is powered by a 6-cylinder diesel engine. The amphibious vehicle has a maximum road speed of 100km/h and a road range of 600km. The vehicle in the picture belongs to D Company of 4th Light Armored Reconnaissance Battalion, USMC.

Among the LAV fleet of 4th Light Armored Reconnaissance Battalion of the USMC that was deployed to Norway, was this Light Armored Recovery Vehicle (LAV-R). The LAV-R has a four-man crew consisting of the driver, commander and two mechanics. For recovery and maintenance operations, the vehicle is fitted with a 1.8-ton hydraulic crane, a 13.6-ton winch, and an auxiliary power unit with power tools. For self-defense purposes, the LAV-R is armed with a pintle-mounted M240 machine gun and two M257 four-barreled smoke grenade launchers. This particular vehicle is a veteran of Operation Iraqi Freedom and has seen active service in Iraq.

During Exercise Battle Griffin 2005, MAGTF-25 was supported by Bv 206 all-terrain vehicles of the Norwegian Army. The vehicles were supplied to the U.S. Marines together with drivers. The Bv 206s shown in the picture are some of these fore mentioned vehicles.

A crewmember of a CV 9030 of a Norwegian reconnaissance platoon is using his binoculars to scan the outskirts of a Ranian enclave. Note the handheld GPS situated in front of him.

The CV 9030N has replaced the NM 135 as the infantry fighting vehicle in the Norwegian Army inventory. The CV 9030N has a crew of three, consisting of a driver, commander and gunner. In the rear compartment up to eight fully equipped infantry men can be transported. The CV 9030N has a combat weight of 26 tons and has an outstanding cross-country capability. The pictured vehicle belongs to Mechanized Infantry Battalion 2, which made up part of the MNTF.

This well-camouflaged CV 9030N belongs to the heavy reconnaissance platoon of Mechanized Infantry Battalion 2. The CV 9030N is armed with a 30mm Bushmaster II cannon. Coaxial to the weapon, a 7.62mm MG3 machine gun is mounted. On board the CV 9030N, 400 rounds of 30mm ammunition and 3,800 rounds of 7.62mm ammunition are carried.

This picture shows a vehicle commander of a Norwegian CV 9030N infantry fighting vehicle.

Mechanized Infantry Battalion 2 has, under its HQ and Support Company, a heavy reconnaissance platoon that is equipped with CV 9030N infantry fighting vehicles. Here the four vehicles of the platoon can be seen in position next to an advance route of the MNTF. Note the vehicle's winter camouflage consisting of camouflage netting and whitewash.

This CV 9030N belongs to the the Badufoss-based "Panser Battaljonen i Nord Norge". Note the vehicle's markings that consist of a sword-wielding griffin and the tactical number "91A".

Norwegian engineers stand guard at their unit's Forward Operating Base (FOB). Due to the "Three Block War" character of the exercise scenario, the participating troops constantly had to be on the alert for terrorist attacks. While the machine gunner on the M113A1 troop carrier is manning a 12.7mm M2HB machine gun, his comrade is armed with a 7.62mm G3 assault rifle with a sliding stock.

This Bv 206 of an engineer unit is one of the flatbed versions. The vehicle has a payload of 2,350kg and its rear trailer can be fitted with a wide range of equipment, in this case a container. In the front cab there is space for the driver and five troops. A weapon ring is mounted on the vehicle's roof, which is here equipped with a 12.7mm M2HB machine gun.

This M113A1 ambulance taking part in Exercise Battle Griffin 2005 is of particular interest. The vehicle was fitted with new tracks, an endless rubber band track reinforced with steel wire. On the rubber track, special snow grousers can be mounted that prevent the vehicle from sliding on icy roads. The new track also reduces vibrations when on the move, and therefore makes traveling much more comfortable for the crew and patients. The track is currently undergoing testing by the Norwegian Army.

During Exercise Battle Griffin 2005, the exercise troops often had to operate in temperatures well below zero. Here, a Norwegian soldier has protected himself against the cold with a ski mask and protective goggles.

The Bv 206 all-terrain vehicle is used in countless models by the Norwegian Army and over 2,200 vehicles are in service. Here the standard troop carrier can be seen towing a trailer. The vehicle is armed with a 12.7mm M2HB machine gun. In the front cab there is space for five soldiers and the driver, while in the rear cab another eleven troops can be carried.

1/35 SCIMITAR

Laurent Lecocq

Flugabwehrbrigade 100

Tim Mätzold, Daniel Nowak, Clemens Niesner

Shortly after having reached its fire position this ROLAND has raised its launchers and radar system. Although trials with a prototype vehicle named ROLAND NDV (Nutzungsdauerverlängerung (Life Extension Programme)) for linking the system to the HflaAFüSys were undertaken in the middle of 2003, the withdrawal of FlaRakPz 1 ROLAND 2 will prevent the ROLAND NDV from entering into service. (Tim Mätzold)

After having analyzed the new political situation in the world, the German government has decided to restructure the Bundeswehr to meet the requirements of the 21st century. In most cases this means a reduction of manpower and material for the armed forces. The future intention of the strategic defense review released on 29th January 2001 was to form five similarly structured mechanized divisions, one division for special operations (Division für Spezielle Operationen (DSO)) and one division for airmobile operations (Division für Luftbewegliche Operationen (DLO)). The combat support troops (of which the FlaBrig100 belongs to) and the logistics troops will be concentrated in brigades within these so-called "Heerestruppenkommando" (command of armed forces).

The result was the establishment of an armored air defense brigade that includes all remaining air defense units enabling them to fulfill these basic tasks:

- Close air defense at low and high altitude flight levels
- Securing objectives and strategically important areas
- To weaken enemy air superiority and to strengthen their own air defense

The FlaBrig 100 is equipped in a way which will support the heavy mechanized divisions of the German ground forces, as well as the light and flexible response forces. The Flugabwehrbrigade 100 (FlaBrig100 [Air Defense Brigade 100]) was finally established on 1st July 2002, having its HQ at Fuldatal/Rothwesten. About 4,100 soldiers and 2,500 reservists of five federal counties are put under the FlaBrig 100's command.

The FlaBrig100's three main weapon systems (Gepard, Roland, and Ozelot) are stylistically portrayed in black and white coloring as the centerpiece of the insignia. The blue color represents the sky and is the base color, while the coral red color stands for the air defense troop.

This soldier guards the staging area of an exchange crew using a 7.62mm MG3 on an improvised gun mount. A special target locating device is used for engaging air targets (this requires an uncanny knack!) (Clemens Niesner)

A gunner of a dismounted air defense unit, equipped with an SA-7 GRAIL 9M32M STRELA-2 "Fliegerfaust" (shoulder-launched air defense system), aims at a target. The average range is between 500 – 5,500m. Infrared-guided, the STRELA is armed with an HE warhead, and can reach a top speed of 580m/sec. The weapon has a length of 1.47m and a weight of 9.97kg. The method of camouflage shows the soldier's high state of readiness when it comes to infantry skills! (Clemens Niesner)

The optical target locating device enables soldiers to survey a second area at a distance of about 50m from the vehicle. Furthermore the weapon system can be controlled by a cable-linked remote control. (Daniel Nowak)

Each air defense troop has its own maintenance section consisting of an MB290GD 0.9t gl WSA (Werkstattausrüstung (Fitters section)). The vehicle tows a trailer fitted with spare parts that should last for 30 days, and special electronic measuring instruments. An interactive electro-mechanical documentation plant (Interaktive Elektronische Technische Dokumentation IETD) supports the technicians in doing their job. (Daniel Nowak)

Structure and equipment

Since the summer of 2004, the following units were put under the FlaBrig100's command:

Stab & Stabsbattery (HQ and HQ battery), Fuldatal
- Flugabwehraufklärungsbatterie (air defense reconnaissance battery) 100, Fuldatal
- Leichte Flugabwehrraketenbatterie (light air defense rocket battery) 100, Borken
- Leichte Flugabwehrraketenbatterie (light air defense rocket battery) 300, Fuldatal
- Leichte Flugabwehrraketenlehrbatterie (light air defense training rocket battery) 610, Lütjenburg
- Panzerflugabwehrkanonenbataillon (armored air defense battalion) 6, Lütjenburg
- Panzerflugabwehrkanonenbataillon (armored air defense battalion) 12, Hardheim
- Panzerflugabwehrkanonenbataillon (armored air defense battalion) 131, Hohenmölsen
- Panzerflugabwehrraketenbataillon (armored rocket air defense battalion) 7, Borken
- Panzerflugabwehrraketenbataillon (armored rocket air defense battalion) 300, Fuldatal

Next to these active units, the following non-active units also belong to the FlaBrig100:

- Panzerflugabwehrkanonenbataillon 61, Lütjenburg
- Panzerflugabwehrkanonenbataillon 121, Hardheim
- Panzerflugabwehrkanonenbataillon 132, Hohenmölsen
- Panzerflugabwehrraketenbataillon 101, Borken
- Panzerflugabwehrraketenbataillon 301, Fuldatal

The FlaBrig100 has the following totals of weapon systems:
4x Luftraumüberwachungsradar (air defense surveillance radar) LÜR
126x Flugabwehrkanonenpanzer (FlaKPz [air defense tank]) GEPARD
100x Flugabwehrraketenpanzer (FlaRakPz [air defense rocket tank]) ROLAND
45x Leichter Flugabwehrraketenträger (light air defense rocket vehicle) OZELOT
26x Flugabwehrführungspanzer (air defense command vehicle) FUCH

Structure & equipment of the Flugabwehraufklärungsbatterie (FlaAufklBtr [Air Defense Reconnaissance Battery])

The FlaBrig100 is currently equipped with the "Luftraumüberwachungsradar LÜR" (air defense surveillance radar). Together with the "Tieffliegerüberwachungsradar TÜR" (radar for low flying aircraft), they are the most modern systems of their kind. It is based on the 15t mil gl KAT1A1 (8x8) truck. The radar can cover a range of 100km and picks up all aircraft within this range. Secondly, enemy aircraft and helicopters can be identified.

The FlaKPz GEPARD has 3 shift-working crews. Whilst one crew relaxes or has a meal, the other crew provides dismounted air-defense; and the third crew serves in the basic task of operating the air defense tank. The exchange crews are transported by a Unimog 2t gl truck and here they join an exercise break at a forest near Göttingen during Exercise "Thüringer Löwe 2004". (Clemens Niesner)

One battery consists of:
- Battery command group
- Two air defense reconnaissance troops with one command troop each, two LÜR, one TÜR, four interface troops (FAST) with a Unimog 2t gl, and three ground troops equipped with shoulder-launched air defense weapons for self-defense
- Technical group
- Support group

Prompted by the withdrawal of the FlaRakPz ROLAND, the scheduled establishment of the air defense command post ROLAND FGR based on the 15t mil gl KAT1 A1 (8x8) truck will be cancelled.

Connection of the "Heeresflugabwehr – Aufklärungs – und Gefechtsfeldsystem (HflaAFüSys)" (air defense reconnaissance and battlefield command system)

Enemy air strikes and air reconnaissance are a major threat to ground forces on the battlefield. HflaAFüSys supports the air defense troop in a way that effectively engages that threat and helps keep ground operations protected. The whole scope of reconnaissance, command and fire command tasks are integrated by this system.

1. Reconnaissance

This task establishes a general air defense network for every single radar sensor, the fire control vehicle, the battery command post and the individual weapon systems. The information about the current air defense situation with allied fire command and weapon systems is done by the FAST interface (Flugabwehr Aufklärungsschnittstelle Tiefflugbereich [air defense reconnaissance interface for low flying aircraft]).

2. Command

The command provides information and functions required for releasing the right orders and making the right decisions:
- situations, reports, orders and fire rules
- flight schedules
- identification friend-foe
- situation and location of friendly units
- air defense areas

Other command and information systems can be linked to their own command circuit.

3. Fire control

This cell analyses the air defense situation to estimate the threat, evaluate the state of the weapon systems (checking the fire rules) and providing the information required for proper target location.

All air defense units equipped with GEPARD or ROLAND use the TPz-1 A5A1 FUCHS command post vehicle fitted with the FlaFü (Flugabwehr Führung (Air Defense Command)) attachment and the HFlaAFüSys (Heeres Flugabwehr Aufklärungs – undFührungssystem (air defense reconnaissance and command system)). This is integrated with the HEROS (Heeres –Führungsinformationssystem (ground forces command information system)) for computer- supported staff operations. This vehicle belongs to "Flugabwehrkanonenbataillon 12" from Hardheim and is participating in Exercise "Thüringer Löwe 2004". (Daniel Nowak)

As soon as the fire control officer knows the current air defense situation, he begins to command his weapon systems from his own fire command vehicle. Every combat vehicle gets its own target and reports back on final destruction of their targets.

The GEPARD has a major advantage as its own target locating radar does not have to be used, since the data is delivered by the HFlaAFüSys. This makes enemy reconnaissance more difficult and helps with survival on the battlefield!

The use of VHF radios meets the high requirements of modern communications:
- mobility
- crypto-capability
- resistance against interference
- transfer safety
- data rate
- range

The reconnaissance and battery radio circuit transfers digital data.

This fire control officer uses the TPz-1 A5A1 FUCHS FlaFü. After having received the general air defense situation, he provides fire control for all weapon systems that have been put under his command. This TPz FUCHS belongs to the staff battery of "Flugabwehrbrigade 100". (Clemens Niesner)

The air defense surveillance radar LÜR/TRMS (Luftraum Überwachungs Radar/Tele-funken, Radar, Mobil, Such) has an average range of 200km. It is the central core of the HflaAFüSys as it gathers and investigates basic information about all aircraft and helicopters in the defense area. The Phased-Array-Radar operates in the C-Band area and is resistant to electronic countermeasures. It is able to track up to 4,000 targets simultaneously, enabling it to establish a general air defense state. (Clemens Niesner)

A ROLAND of PzFlaRakBtl300 negotiates its way through a small village near Warburg – a sight that will soon be consigned to the past. This is because the ROLAND air defense system is supposed to be withdrawn as it is not linked to the HFLaAFüSys. (Clemens Niesner)

This shot shows a FlaKPz GEPARD 1A2 fitted with a large additional stowage bin on the vehicle's rear. The radar system is lowered for road transport. Next to the main engine the GEPARD has an auxiliary engine providing power for the weaponry and electrical systems. (Clemens Niesner)

The rocket launchers of this ROLAND belonging to Panzerflugabwehrraketenbataillon 300 from Fuldatal are lowered for road marches. Both the ROLAND and the GEPARD have their own auxiliary engines providing power for the weapon and electrical systems. This photograph was taken on Exercise "Leine Express 2" in March 2001 near Moringen. (Tim Mätzold)

Leichtes Flugabwehrsystem LeFlaSys (Short Range Air Defense System SHORAD)

SHORAD is currently the most modern air defense system within the FlaBrig100. This air-portable air defense system is named OZELOT and is based on the chassis of the WIESEL 2 light air-portable tracked vehicle. It is designed to engage enemy aircraft, helicopters and rockets.

OZELOT was developed by STN ATLAS in 2002, though Krauss-Maffei-Wegmann (KMW) was involved in the project later on. The first TVM (Truppenversuchsmuster [prototype vehicle]) left the plant in 1996. Two years later the LeFlaSys began its official production.

The Heeresflugabwehrschule (school of air defense) at the Feldwebel Schmidt Kaserne, Rendsburg, received the very first vehicle on 27th June 2001. All light air defense batteries are currently equipped with the LeFlaSys. Other countries are interested in the new system as well. Greece has ordered some 54 Atlas Short Range Air Defense Systems (ASRAD – Hellas) based on HMMWVs from 2005 on. The Finnish Army has ordered the system based on a Unimog U5000L; and even the Netherlands Army will sign a contract shortly.

The OZELOT is equipped with the ASRAD Pedestal guided weapon system that is armed with 4 AN/FIM-92 Stinger rockets. Having an average flight level of 2,000m, the system can engage targets at a distance of up to 6,000m. The HE warhead can destroy enemy aircraft and helicopters with a first hit. The vehicle is fitted with special storage boxes at the rear for storing another 4 rockets, meaning a fully laden vehicle is armed with 8 Stinger rockets. Next to the Stinger and SA-16 IGLA, the system can also use MISTRAL, STARBURST and RBS70 Mk2 air defense missiles as well!

The weapon systems can be traversed 360° and are lowered for road transport. It is fitted with a sophisticated target locating system which consists of the following parts:

- Sensor and electronic unit equipped with the Zeiss Ophelius thermal sight
- All-weather target locating device
- High intensity CCD camera
- Eye-safe laser rangefinder MOLEM

In emergency cases the rockets can be launched manually.
The system is highly efficient as the FlaFü WIESEL 2 (command vehicle) and WIESEL 2 AFF (Aufklärungs, Führungs - und Feuerleitfahrzeug [reconnaissance, command and fire control vehicle]) components work together with the OZELOT air defense vehicles. At the troop or battery level, all OZELOTs receive their target data from the WIESEL 2 AFF via digital radios.

A SHORAD (Short Range Air Defense System) "OZELOT" in a somewhat concealed position. This version fitted with sealed launch containers means only STINGER guided missiles can be launched. The system can work autonomously without the WIESEL 2 AFF using the Thales IR infrared surveillance system (IRST). The armor provides protection against 7.62mm rounds and splinters from 155mm artillery shells at a distance of 20m. (Daniel Nowak)

Up to a range of 20km the defense area is searched autonomously by the 3D HARD Radar (Helicopter & Airplane Radio Detection). A WIESEL 2 AFF has the capability of providing fire control for up to 8 OZELOTs. In the case of an AFF system failure, OZELOT is able to track targets on its own using the IRST passive Infrared Search and Track sensor (there is one IRST per air defense troop).

The FlaFü WIESEL 2 gathers reports and information to coordinate all air defense troops. The AFF WIESEL 2, FlaFü WIESEL 2 and 0,9t FAST truck have one digital radio link using the SEM93 and SEM90 radios. Furthermore, the LeFlaSys can use information provided by the Heeresflugabwehr Aufklärungs – und Führungssystem HFlaAFüSys.

Besides the armed launchers which carry 4 LFK AN/FIM 92A STINGERs, a further 4 guided missiles can be stored in special ammunition cases within the vehicle. For engaging enemy aircraft the commander receives the target data from the AFF WIESEL 2 by digital radio. The weapon system tracks the target automatically by just pushing a button before it finally destroys it. Up to 2 fire missions can be done at the same time. It is not necessary to identify the enemy aircraft again as this has been done by the AFF already. (Daniel Nowak)

Every light air defense battery consists of:
15x WIESEL 2 OZELOT Short Range Air Defense Systems
3x WIESEL 2 AFF command and reconnaissance vehicles
2x WIESEL 2 FlaFlü Fire control and command vehicles
1x 0,9t gl MB Wolf FAST truck (interface low flying aircraft)
3x 0,9t gl LL truck (air-portable) with fitters section.
1x LKW 0,9t gl MB WOLF vehicle
1x Unimog 2t gl truck carrying exchange crews equipped with shoulder-launched air defense weapons.

Flugabwehrkanonenpanzer (FlaKPz [Air Defense Tank]) GEPARD

The FlaKPz GEPARD, which is based on the Leopard 1 chassis, is currently the heaviest air defense system of the FlaBrig100. It began development in 1965 and produced from 1976 onwards.

By 1980 the German Army had received some 420 FlaKPz in different versions that had been built by Krauss-Maffei Wehrtechnik. The turret was produced by Wegmann, the fire control computer is a Siemens product and the gun systems were made by the Swiss enterprise Oerlikon-Contraves. The two 35mm Oerlikon KDA fast fire guns are the GEPARD's main weapon system and can fire about 550 rounds per minute with a VO (Muzzle Velocity) of 1,175m/sec. 660 rounds can be stored within the vehicle.

The radar mounted at the turret's rear can cover a range of 15km and is lowered for road marches. All aircraft are identified by IFF (Identification Friend-Foe). After having located an enemy aircraft the tracking radar mounted on the front of the turret switches on. The fire control system consists of a tracking radar, an electro-mechanical fire control computer, the VO measuring device, and other optical tracking systems. The fire control computer uses target location data (range, velocity and flight screen) for estimating the exact destruction location and starts firing about 6 seconds after having tracked the target. It has an average range of 3,500m. The German Army has given 43 GEPARDs to Romania for supporting this new NATO member.

Between 1997 and 2000, 148 GEPARD air defense tanks underwent a life extension program to upgrade them to an A2 standard (the 191 older vehicles of B2L standard were withdrawn). The A2 standard includes a

An AFF WIESEL can provide target locating data for 5 – 9 SHORADs using the SEM 93 digital radio. Additionally, other allied air defense information from other systems can be evaluated. The HARD 3D radar can cover a range and height of 20km. It uses an MTI (Moving Target Indication) and an MSR 200 XE IFF (Identification Friend-Foe) system. (Daniel Nowak)

new digital fire control computer that replaces the old one, a cooling device for turret compartments and improved ammunition – FAPDS. The new SEM93 radios enable a connection to the "Heeres-Flugabwehr Aufklärungs - und Führungssystem (HflaAFüSys)".

An armored air defense regiment consists of an HQ battery (including all elements of command, support and maintenance), four armored air defense batteries with six FlaKPz GEPARDs each (including support vehicles), the FUCHS wheeled command post vehicle and a Unimog 2t gl truck for the reserve crews.

ROLAND rocket air defense system

Next to the GEPARD armored air defense tanks, the ROLAND rocket air defense system is employed within the "Flugabwehr Rakentenbataillone" (air defense rocket regiments). It is based on the MARDER infantry fighting vehicle.

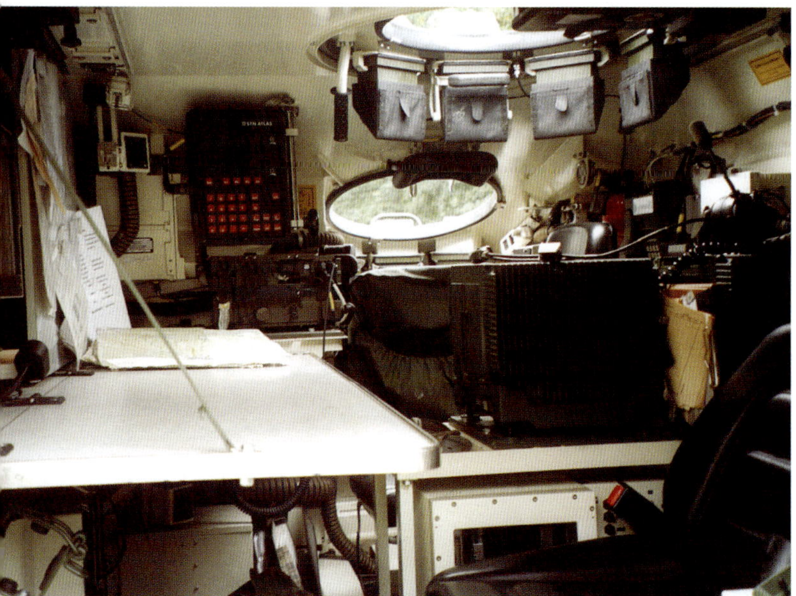
This shot shows a WIESEL 2 BF/UF (Unterstützungs – und Batterie – Führungszelle Flugabwehr (Support and Battery Command Cell Air Defense)) from the inside. The vehicle is the mobile command post of a light air defense battery. A crew of 3 soldiers evaluates all troop reports, coordinates them and gives orders. Apart from the SEM93 digital radio and the SEM 90 radio, the WIESEL BF/UF is also fitted with the HEROS. (Archiv Daniel Nowak)

Even in a time of modern technology, close air defense still plays an important role on the battlefield! A soldier of a dismounted air defense troop surveys his area with a pair of binoculars from the former NVA (Nationale Volksarmee (Eastern German Army)). (Archiv Daniel Nowak)

In 1962 a new air defense system was developed as a joint venture between France, Germany and Great Britain. Whilst Great Britain eventually started its own project, Germany and France kept on working on the future air defense system; resulting in the first prototype vehicle based on the RU242 chassis. Three years later the first ROLAND rocket (manufactured by Euromissile) was launched. Final production of regular vehicles started on 15th June 1981 at Kassel.

A total of 143 vehicles (including three troop trial vehicles) were produced and delivered to three air defense units (at the corps level) and one "Heeresflugabwehrbataillon" (air defense regiment) of 6.Panzergrenadierdivision. The somewhat prolonged development time was prompted by the joint venture between Germany and France, as Germany decided to develop the all-weather ROLAND2 version. Even the U.S. Army was interested in manufacturing the system under license but its bid failed through a lack of money! Brazil also uses 4 ROLAND FlaRakPz (Flugabwehr Raketen Panzer [Rocket Air Defense Systems]) alongside Germany and France.

The weapon system is located in the vehicle's rear compartment and consists of two rocket launchers, each with one LFK (Lenkflugkörper [guided missile]) ROLAND 2. The rocket is armed with a warhead carrying 6.5kg of TNT, enabling it to destroy targets at a range of about 6,000m and at an average height of 4,000m, and at a speed of Mach 1.5. A total of 10 rockets can be stored within the vehicle - 8 rockets are carried in special storage boxes inside the vehicle while two remain ready in the launchers.

The electronic devices consist of a surveillance radar (2D) with Identification Friend-Foe (IFF), infrared goniometer, command computer, command transceiver and a 3D tracking radar. A change to optical tracking is possible during the procedure. As ROLAND will be withdrawn in the future it will not be upgraded so as to be linked to the HFlaAFüSys.

Both air defense rocket regiments consist of an HQ battery and four firing batteries (2 batteries in reaction forces, and 2 batteries in reinforcing troops) with 8 ROLANDs each.
Similar to the armored air defense regiments, the FUCHS wheeled APC is used as a command vehicle whilst the UNIMOG 2t gl carries the second crew.

Future Preview

The current strategic defense review dated 1st November 2004 will restructure the army concerning their new tasks of worldwide operational

This FlaRakPz 1 ROLAND 2 was camouflaged with chalk to help blend it in with the wintry environment. Note the weapon system with the raised oval Thomson CFS-D Band tracking radar and the Siemens search radar with a range of approximately 16km, and a height range of 4km. (Clemens Niesner)

This shot shows the TPz FUCHS FüFuFlaFü – FlgLt (Führungsfunk für Flugabwehrführung (command radio for command air defense)) of 3.Bttr./PzFlaKBtl. 131 from Hohenmölsen. German air defense uses this version within armored air defense and rocket air defense regiments. (Clemens Niesner)

(peacekeeping) missions. Finally the ROLAND system will not be used anymore as their specialist tasks will likely be fulfilled by PATRIOT units. With the reduction in size of the German Army, the German air defense forces will be reduced to 60% of their former strength, prompted by the withdrawal of the "Flugabwehrbrigade 100". The remaining units will be the "Panzerflugabwehrkanonen Lehrbataillon (PzFlaKLBtl [Armored Air Defense Training Regiment]) 6", "Panzerflugabwehrkanonentabaillon 12", "Leichte Flugabwehr Raketenbatterie 300" (le PzFlaRakBtr300 [Light air defense rocket battery])", LeFlaRakBtr100 and the "Leichte Flugabwehr Raketen Lehrbatterie (LeFlaRakLBtr) 610 [Light Air Defense Rocket Training Battery])".

To supply the sophisticated radio systems with electrical power, the TPz-1 FüFu and FlaFü are equipped with an auxiliary power unit (SEA (Strom Erzeuger Aggregat)) equipped with a VW engine delivering 5 kW. It is situated in a position on the left rear door. (Clemens Niesner)

Technical data

	FlaRakWaTrg Ozelot	**FlaKPz Gepard**	**FlaRakPz Roland**
Crew:	2 Men	3 Men	3 Men
Length:	4.50m	8.15m	7.13m
Width:	1.82m	3.20m	3.25m
Height:	2.55m	3.03m	4.63m
Combat Weight:	4.07t	47.38t	32.5t
Power to Weight Ratio:	27.02PS/t	17.5PS/t	18.46PS/t
Top Speed:	70km/h	65km/h	70km/h
Climbing Capability:	60%	60%	60%
Fording Depth:	0.50m	1.20m	1.50m
Engine:	VW 1,9l Tdi 4 Cylinder Supercharged Diesel Fuel engine with direct injection and intercooler	MTU MB838 Ca500M V10 Cylinder super-charged multifuel engine	MTU MB833 EA500 V6 Cylinder super-charged multifuel engine
Engine Power :	81kw/110PS	610kw/830PS	441kw/600PS
Gearbox:	ZF 4-gear automatic with torque converter 4 HP24	4 Forward/ 2 reverse gears	4 forward/ 2 reverse gears
Fuel Tank:	120l	985l	652l
Range :	550km	550km (road) 400km off-road	520km (road) 200km off-road
Armament:	4x LFK AN/FIM-92 Stinger 7.62mm MG3	2x 35mm chain gun OERLIKON KDA 2 x 4 smoke dischargers	2 LFK Roland 2 1 Fla – MG 7.62mm 2 x 4 smoke dischargers
Manufacturer:	Krauss Maffei Wegmann GmbH, München STN ATLAS Elektronik GmbH, Bremen, Germany	Krauss-Maffei Wegmann München, Germany Oerlikon-Contraves, Zürich, Swizerland	Thyssen-Henschel, Kassel, Germany

Soldiers of the light air defense units are able to carry out airdrop operations with the STINGER shoulder-launched air defense weapon. Together with other allied paratroops they are able to form a first limited air defense belt with MANPADS (Man Portable Air Defense Systems) to secure the proceeding of SHORAD. (Clemens Niesner)

The German air defense's 3 main weapon systems pictured together. A battle of combined arms, including the capabilities of conventional and rocket artillery, can achieve high efficiency. All armored air defense vehicles are additionally armed with the "Fliegerfaust STINGER" or "STRELA", both shoulder-launched air defense weapons. (Clemens Niesner)

From 1988 on, a total of 206 FlaKPz 1 GEPARDs underwent the first phase of a life extension program by getting a laser rangefinder. A total of 640 rounds of FAPDS (air defense ammunition) and 40 rounds of HVAPDS-T (High Velocity Armor Piercing Discarding Sabot – Tracer) for engaging ground troops can be stored within the vehicle. The armor piercing ammunition can destroy ground targets effectively at a range of up to 1,200m. (Clemens Niesner)

This shot shows clearly the sensor compartment between the 2 launchers. The compartment consists of a Zeiss OPHELIUS thermal sight, a TV camera, a laser rangefinder (with a range of about 10km) and a heavy-duty tracker. Furthermore, the vehicle is fitted with the TALIN 3000 hybrid navigation system by Honeywell ,a GPS receiver by Rockwell Collins and 2 SEM 90 and 93 VHF radios. (Clemens Niesner)

Both the FlaKPz GEPARD and the FlaRakPz ROLAND are autonomously working all-weather air defense weapons. Due to its high mobility and heavy armor, the GEPARD is well designed to protect mechanized troops from enemy air strikes. (Clemens Niesner)

The "Gefechtsübungszentrum Heer" (GefÜbZ H) north of Magdeburg provides formidable possibilities to train a combined arms battle under the most realistic of conditions! A combination of simulation technology, consisting of the AGDUS duel simulation (Direct Fire Weapon Effect System) and the "Wirkungsdarstellung" (WD) (effect simulation) with green/yellow flashing beacons, puts every crew through their paces. This GEPARD 1A2 is fitted with the complete simulation kit. (Clemens Niesner)

With a rate of fire of 2x500 rounds per minute, a muzzle velocity of 1,400m/second and the use of high-explosive FADPS rounds, it is likely that a target can be destroyed with a first hit! The GEPARD 1A2 will get C3 capabilities (Command, Control and Communication) when linked to the HflaAFüSys. (Clemens Niesner)

This GEPARD 1A2 of PzFlaKBtl 12 crosses a "Hohlplattenbrücke" (hollow plate bridge) over the Haves River near Rathenow during Exercise "Thüringer Löwe 2003". Note the barrels of the 35mm Oerlikon KDA machine guns. (Clemens Niesner)

After having launched the ROLAND 2 guided weapon, the launch container is withdrawn; the launcher arm takes a new rocket out of the magazine and reloads the launcher. The whole procedure takes about 6 seconds. Up to 8 rockets can be stored within the vehicle. (Clemens Niesner)

After having reached a firing position, the crew camouflages the vehicle immediately and the Siemens Pulsdoppler-radar goes into action. The whole vehicle will be camouflaged with the exception of the weapon system. (Clemens Niesner)

A fitters section, equipped with a Unimog 1300L 2t gl (4x4) and a Bergepanzer 2A2 Standard armored repair and recovery vehicle, provides technical support. They belong to the HQ battery. (Clemens Niesner)

A FlaRakPz ROLAND shortly before moving to a new firing position during Exercise "Thüringer Löwe 2004". The vehicle has a crew of 3 soldiers: there is the driver, a commander (who commands the vehicle and the fire mission), and a gunner (he is not seen in this shot) who is responsible for operating the weapon system and destroying the target. (Clemens Niesner)

Besides the Unimog 2t gl, the MAN LKW 5t mil gl KAT 1 is used within the air defense regiments. In this case it is the 2.Batterie, Panzerflugabwehrkanonenbataillon 12 from Hardheim. This vehicle belongs to the troop of the battery sergeant-major. (Clemens Niesner)

Both the GEPARD and the ROLAND regiments use the Bergepanzer 2A2 Standard ARRV. Apart from its other features, the A2 version has a hydraulic support at the rear of the vehicle for better stability when using the crane. The German Army has more than 100 of these upgraded ARRVs. (Clemens Niesner)